Lecture Notes in Artificial Intell

Edited by J. G. Carbonell and J. Siekmann

T0238169

Subseries of Lecture Notes in Computer Science

Katharina Morik Jean-François Boulicaut
Arno Siebes (Eds.)

Local Pattern
Detection

International Seminar
Dagstuhl Castle, Germany, April 12-16, 2004
Revised Selected Papers

 Springer

Series Editors

Jaime G. Carbonell, Carnegie Mellon University, Pittsburgh, PA, USA
Jörg Siekmann, University of Saarland, Saarbrücken, Germany

Volume Editors

Katharina Morik
University of Dortmund, Computer Science Department, LS VIII
44221 Dortmund, Germany
E-mail: morik@ls8.cs.uni-dortmund.de

Jean-François Boulicaut
INSA Lyon, LIRIS CNRS UMR 5205
Batiment Blaise Pascal
69621 Villeurbanne, France
E-mail: Jean-Francois.Boulicaut@insa-lyon.fr

Arno Siebes
Utrecht University
Department of Information and Computing Sciences
PO Box 80.089, 3508TB Utrecht, The Netherlands
E-mail: arno.siebes@cs.uu.nl

Library of Congress Control Number: 2005929338

CR Subject Classification (1998): I.2, H.2.8, F.2.2, E.5, G.3, H.3

ISSN   0302-9743
ISBN-10   3-540-26543-0 Springer Berlin Heidelberg New York
ISBN-13   978-3-540-26543-6 Springer Berlin Heidelberg New York

Springer is a part of Springer Science+Business Media

springeronline.com

© Springer-Verlag Berlin Heidelberg 2005
Printed in Germany

Typesetting: Camera-ready by author, data conversion by Boller Mediendesign
Printed on acid-free paper   SPIN: 11504245   06/3142   5 4 3 2 1 0

Preface

Introduction

The dramatic increase in available computer storage capacity over the last 10 years has led to the creation of very large databases of scientific and commercial information. The need to analyze these masses of data has led to the evolution of the new field knowledge discovery in databases (KDD) at the intersection of machine learning, statistics and database technology. Being interdisciplinary by nature, the field offers the opportunity to combine the expertise of different fields into a common objective. Moreover, within each field diverse methods have been developed and justified with respect to different quality criteria. We have to investigate how these methods can contribute to solving the problem of KDD.

Traditionally, KDD was seeking to find global models for the data that explain most of the instances of the database and describe the general structure of the data. Examples are statistical time series models, cluster models, logic programs with high coverage or classification models like decision trees or linear decision functions. In practice, though, the use of these models often is very limited, because global models tend to find only the obvious patterns in the data, which domain experts already are aware of[1]. What is really of interest to the users are the local patterns that deviate from the already-known background knowledge. David Hand, who organized a workshop in 2002, proposed the new field of local patterns.

The Dagstuhl Seminar in April 2004 on Local Pattern Detection brought together experts from Europe, Japan, and the United States – 13 countries were represented. Moreover, the participants brought with them expertise in the following fields: decision trees, regression methods, bayesian models, kernel methods, inductive logic programming, deductive databases, constraint propagation, time series analysis, query optimization, outlier detection, frequent set mining, and subgroup detection. All talks were focused on the topic of local patterns in order to come to a clearer view of this new field.

Novelty of Local Pattern Detection

Researchers have investigated global models for a long time in statistics and machine learning. The database community has inspected the storage and retrieval of very large datasets. When statistical methods encounter the extremely large amount of records and the high dimensionality of the stored observations, exploratory methods failed. Machine learning already scales up to build up global

---

[1] I. Guyon, N. Matic and V. Vapnik. Discovering informative patterns and data cleaning. In *Advances in Knowledge Discovery and Data Mining* (pp. 181–204). AAAI Press/MIT Press, 1996.

models, either in the form of complete decision functions or in the form of learning all valid rules from databases. However, the classification does not deliver new, surprising insights into the data, and the valid rules reflect almost exactly the domain knowledge of the database designers. In contrast, what users expect from the exploratory analysis of databases are new insights into their data. Hence, the matter of interestingness has become a key issue. The success of Apriori or subsequently frequent set mining can be explained by it being the first step into the direction of local patterns. The correlation of more than the few features, which standard statistics could analyze, could successfully be determined by frequent set mining. Frequent set mining already outputs local patterns. Current research tasks within this set of methods include algorithmic concerns as well as the issues of interestingness measures and redundancy prevention. The collaboration of database specialists and data miners has led to the notion of inductive databases. The new approach writes measures of interest and the prevention of redundancy in terms of constraints. Also users can formulate their interests in terms of constraints. The constraints are pushed into the search process. This new approach was discussed at the seminar intensively and a view was found that covered diverse aspects of local patterns, namely their internal structure and the subjective part of interestingness as given by users.

Not all the exciting talks and contributions made their way into this book, particularly when a version of the talk was published elsewhere:

- Rosa Meo presented a language for inductive queries expressing constraints in the framework of frequent set mining.
- Bart Goethals offered a new constraint on the patterns, namely that of the database containing the minimal number of tiles, where each tile has the maximal number of '1'.
- Stefan Wrobel gave an in-depth talk on subgroup discovery, where he clearly indicated the problem of false discoveries and presented two approaches: the MIDOS algorithm, which finds subgroups according to the true deviation, and a sequential sampling algorithm, GSS, which makes subgroup discovery fast. He also tackled the redundancy problem by maximum entropy suppression effectively. Applications on spatial subgroup discovery concluded the talk.
- Arno Siebes employed a graphical view on data and patterns to express this internal structure. Moreover, aggregate functions along paths in these graphs were used to compute new features.
- Helena Ahonen-Myka gave an overview of sequence discovery with a focus on applications on text.
- Xiaohui Liu explained how to build a noise model using supervised machine learning methods and detect local patterns on this basis. Testing them against the noise model yields clean data. The approach was illustrated with two biomedical applications.
- Thorsten Joachims investigated internal structures such as parse-trees and co-reference pairing. He presented a general method for how such structures can be analyzed by SVMs. Moreover he showed how the combinatorial ex-

plosion of the number of constraints can be controlled by the upper bounds derived from statistical learning theory.

The book then covers frequent set mining in the following chapters:

- Francesco Bonchi and Fosca Giannotti show the use of constraints within the search for local patterns.
- Jean-Francois Boulicaut applies frequent set mining to gene expression data by exploiting Galois operators and mining bi-sets, which link situations and genes.
- Cline Rouveirol reports on the combination of frequent sets found in gene expression and genome alteration data.

Subgroup discovery is represented by three chapters:

- Nada Lavrac reports on successful applications of subgroup mining in medicine.
- Josef Fürnkranz presents a unifying view of diverse evaluation measures.
- Einoshin Suzuki investigates evaluation measures in order to distinguish local patterns from noise.
- Martin Scholz identifies global models with prior knowledge and local patterns with further, unexpected regularities. His subgroup discovery exploits iteratively a knowledge-based sampling method.

The statistical view is presented in the following chapters:

- Niall Adams and David Hand distinguish two stages in pattern discovery
  1. identify potential patterns (given a suitable definition);
  2. among these, identify significant (in some sense) patterns (expert or automatic).
  They notice that the former is primarily algorithmic and the latter has the potential to be statistical. They illustrate this with an application on discovering cheating students.
- Frank Höppner discusses the similarities and differences between clustering and pattern discovery. In particular he shows how interesting patterns can be found by the clever use of a hierarchical clustering algorithm.
- Stefan Rüping introduces a general framework in which local patterns being produced by different processes are identified using a hidden variable. This allows for the use of the EM algorithm to discover the local patterns directly, that is, without reference to the global data distribution. A new scaling algorithm handles the combination of classifiers. The method is illustrated using business cycle data.

Phenomena of time have always been of interest in KDD, ranging from time series analysis to episode learning. Here, three chapters are devoted to time phenomena:

- Claus Weihs focuses on the transformation of local patterns into global models illustrated with the transcription of vocal time series into sheet music.

- Katharina Morik discusses the importance of the example representation, because it determines the applicability of methods. For local pattern detection, frequency features are well suited. She shows how to characterize time-stamped data using a frequency model.
- Myra Spiliopoulou gives an overview of local patterns exhibiting temporal structures, namely changes of (learned) concepts.

Seminar Results

Based on the definition of David Hand[2]

    data = background model + local patterns + random

seminar participants came up with 12 definitions of what local patterns actually are. These were intensively discussed and we finally agreed on the following:

- Local patterns cover small parts of the data space. If the learning result is considered a function, then global models are a complete function, whereas local patterns are partial.
- Local patterns deviate from the distribution of the population of which they are part. This can be done iteratively — a local pattern can be considered the overall population and deviating parts of it are then determined.
- Local patterns show some internal structure. For example, correlations of features, temporal or spatial ordering attributes, and sequences tie together instances of a local pattern.

Local patterns pose very difficult mining tasks:

- Interestingness measures differ from standard criteria for global models.
- Deviation from background knowledge (global model) requires good estimates of the global mode, where local patterns deviate from the overall distribution.
- Modeling noise (for data cleaning, distinguished from local patterns).
- Automatic feature generation and selection for local patterns (for local patterns other features are more successful than for global models; standard feature selection does not work).
- Internal structures of the patterns (correlations of several features, graphs, sequences, spatial closeness, shapes) can be expressed in several ways, e.g., TCat, constraints.
- Test theory for an extremely large space of possible hypotheses (large sets are less likely, hence global models do not encounter this problem).
- Curse of exponentiality — complexity issues.
- Redundancy of learned patterns.
- Sampling for local patterns speeds up mining and enhances quality of patterns.

---

[2] David Hand. Pattern Detection and Discovery. In David Hand, Niall Adams and Richard Bolton, editors, *Pattern Detection and Discovery*, Springer, 2002.

− Evaluation: benchmark missing.
− Algorithm issues.

We hope that this books reflects the issues of local pattern detection and inspires more research and applications in this exciting field.

Katharina Morik
Arno Siebes
Jean-Francois Boulicaut

Table of Contents

Pushing Constraints to Detect Local Patterns 1
Francesco Bonchi, Fosca Giannotti

From Local to Global Patterns: Evaluation Issues in Rule Learning
Algorithms 20
Johannes Fürnkranz

Pattern Discovery Tools for Detecting Cheating in Student Coursework .. 39
David J. Hand, Niall M. Adams, Nick A. Heard

Local Pattern Detection and Clustering 53
Frank Höppner

Local Patterns: Theory and Practice of Constraint-Based Relational
Subgroup Discovery 71
Nada Lavrač, Filip Železný, Sašo Džeroski

Visualizing Very Large Graphs Using Clustering Neighborhoods 89
Dunja Mladenic, Marko Grobelnik

Features for Learning Local Patterns in Time-Stamped Data 98
Katharina Morik, Hanna Köpcke

Boolean Property Encoding for Local Set Pattern Discovery:
An Application to Gene Expression Data Analysis 115
Ruggero G. Pensa, Jean-François Boulicaut

Local Pattern Discovery in Array-CGH Data 135
Céline Rouveirol, Francois Radvanyi

Learning with Local Models 153
Stefan Rüping

Knowledge-Based Sampling for Subgroup Discovery 171
Martin Scholz

Temporal Evolution and Local Patterns 190
Myra Spiliopoulou, Steffan Baron

Undirected Exception Rule Discovery as Local Pattern Detection 207
Einoshin Suzuki

From Local to Global Analysis of Music Time Series 217
Claus Weihs, Uwe Ligges

Author Index 233

Pushing Constraints to Detect Local Patterns

Francesco Bonchi and Fosca Giannotti

Pisa KDD Laboratory
ISTI - CNR, Area della Ricerca di Pisa, Via Giuseppe Moruzzi, 1 - 56124 Pisa, Italy
{francesco.bonchi, fosca.giannotti}@isti.cnr.it

Abstract. The main position of this paper is that constraints can be a very useful tool in the search for local patterns. The justification for our position is twofold. On one hand, pushing constraints makes feasible the computation of frequent patterns at very low frequency levels, which is where local patterns are. On the other hand constraints can be exploited to guide the search for those patterns showing deviating, surprising characteristics. We first review the many definitions of local patterns. This review leads us to justify our position. We then provide a survey of techniques for pushing constraint into the frequent pattern computation.

1 Introduction

The collection of large electronic databases of scientific and commercial information has led to a dramatic growth of interest in methods for discovering structures in such databases. These methods often go under the general name of *data mining*. However, recently two different kinds of structures sought in data mining have been identified: *global models* and *local patterns*.

Traditionally, research in statistics and machine learning has investigated methods to build *global models*, i.e. high level descriptive summarizations of the general structure of the data. Examples are statistical time series models, cluster models, logic programs with high coverage or classification models like decision trees, or liner regression. The intrinsic global nature turned out to be the main drawback that these methods face in practical applications. Having a global point of view over the data, these methods rarely produce new and surprising insight: in fact, in order to be valid, they must summarize most of the data and thus they usually represent obvious knowledge that domain experts are already aware of. On the contrary, what we are seeking for is interesting and surprising knowledge which *deviates* from the already known background model.

Therefore the detection of *local patterns* [14,20] has recently emerged as a new research field with its own distinguished role within data mining. Local patterns are small configurations of data which may involve just a few points or variables, and which are of special interest because they exhibit a deviating behavior w.r.t. the underlying global model. The new field of *local pattern detection* has been proposed by Hand who organized a workshop in 2002 [15]. Such initiative gathered together researchers active in different fields (ranging from statistics to multi-relational data mining, from machine learning to inductive databases) but sharing a common interest for local patterns.

K. Morik et al. (Eds.): Local Pattern Detection, LNAI 3539, pp. 1–19, 2005.

In spring 2004, following the first successful workshop, a Dagstuhl seminar has been organized with the declared goal of finding a definition of local patterns on which most of the participants agree. The lively discussion has produced many slightly different definitions.

1.1   Frequent Pattern Discovery

Even if a rigorous definition of local pattern is still missing, many recognize the successful idea of Apriori [1,2] as a first step into the direction of local patterns. During the last decade a lot of researchers have focussed their (mainly methodological and algorithmic) investigations on the computational problem of *Frequent Pattern Discovery*, i.e. mining patterns which satisfy a user-defined minimum threshold of frequency [2,13].

The simplest form of a frequent pattern is the frequent itemset.

Definition 1 (Frequent Itemset Mining). *Let $\mathcal{I} = \{x_1, ..., x_n\}$ be a set of distinct literals, usually called* items, *where an item is an object with some pre-defined attributes (e.g., price, type, etc.). An* itemset X *is a non-empty subset of $\mathcal{I}$. If $|X| = k$ then X is called a k-itemset. A transaction database $\mathcal{D}$ is a bag of itemsets $t \in 2^{\mathcal{I}}$, usually called* transactions. *The* support *of an itemset X in database $\mathcal{D}$, denoted $supp_{\mathcal{D}}(X)$, is the number of transactions which are superset of X. Given a user-defined* minimum support σ, *an itemset X is called* frequent *in $\mathcal{D}$ if $supp_{\mathcal{D}}(X) \geq \sigma$. This defines the minimum frequency constraint: $\mathcal{C}_{freq[\mathcal{D},\sigma]}(X) \Leftrightarrow supp_{\mathcal{D}}(X) \geq \sigma$. When the dataset and the minimum support are clear from the context, we indicate the frequency constraint simply $\mathcal{C}_{freq}$.*

This computational problem is at the basis of the well known *Association Rules* mining. An association rule is an expression $X \Rightarrow Y$ where X and Y are two itemsets. The association rule is said to be *valid* if the support of the itemset $X \cup Y$ is greater than a given threshold, and if the *confidence* (or accuracy) of the rule, defined as the conditional probability $P(Y \mid X)$, is greater than a given threshold. However frequent itemsets are meaningful not only in the context of association rules mining: they can be used as basic element in many other kind of analysis, ranging from classification [18,19] to clustering [26,29].

Recently the research community has turned its attention to more complex kinds of frequent patterns extracted from more structured data: sequences, trees, and graphs. All these different kinds of pattern have different peculiarities and application fields, (i.e. sequences are particular well suited for *business applications*, frequent subtrees can be mined from a set of *XML documents*, and frequent substructures from graphs can be useful, for instance, in *biological applications*, in *drug design* and in *Web-mining*), but they all share the same computational aspects: a usually very large input, an exponential search space, and a too large solution set. This situation – too many data yielding too many patterns – is harmful for two reasons. First, performance degrades: mining generally becomes inefficient or, often, simply unfeasible. Second, the identification of the fragments of interesting knowledge, blurred within a huge quantity of mostly useless patterns, is difficult. Therefore, the paradigm of *constraint-based mining* was introduced. Constraints provide focus on the interesting knowledge, thus reducing the number of patterns extracted to those of potential interest. Additionally, they can be pushed deep inside the mining algorithm in order to achieve better performance [3,4,5,6,7,8,9,11,12,16,17,21,24,25,28].

Definition 2 (Constrained Frequent Itemset Mining). *A constraint on itemsets is a function* $C : 2^{\mathcal{I}} \rightarrow \{true, false\}$. *We say that an itemset* I *satisfies a constraint if and only if* $C(I) = true$. *We define the theory of a constraint as the set of itemsets which satisfy the constraint:* $Th(C) = \{X \in 2^{\mathcal{I}} \mid C(X)\}$. *Thus with this notation, the* frequent *itemsets mining problem* requires to compute *the set of all frequent itemsets* $Th(C_{freq[\mathcal{D},\sigma]})$. *In general, given a conjunction of constraints* C *the* constrained frequent itemsets mining problem *requires to compute* $Th(C_{freq}) \cap Th(C)$.

In this paper we argue that constraints can be exploited in order to guide the search for local patterns.

2 On Locality, Frequency, Deviation and Constraints

Extending a classical statistical modelling perspective with local patterns, Hand provided the following definition [14]:

$$data = background\_model + local\_patterns + random\_component$$

Based on such definition the participants of the 2004 Dagstuhl seminar discussed in order to find out what a local pattern precisely is.

At the end the participants agreed at least on the following features that a structure must exhibit in order to be considered a local pattern.

1. local patterns cover small parts of the data space;
2. local patterns deviate from the distribution of the population of which they are part;
3. local patterns show some internal structure.

In this Section we will review some of the definitions provided during the seminar by the point of view of frequent pattern mining, arriving to justify our main position: pushing constraints is an important technique to detect local patterns. Hence, let us focus on frequent patterns and association rules. We start our investigation with the obvious question:

Can association rules be considered local patterns?

The answer is not straightforward. On one hand, if we mine the complete set of valid (w.r.t. reasonable support and confidence thresholds) association rules what we obtain is a global descriptive summarization of the data, which for instance, could be used also for classification purposes [19]. Hence what we get is a global model. On the other hand, we could take in consideration a single association rule: is this a local pattern? An association rule is a simple descriptive structure which is true for a reasonably large fraction of the data. It can be understood in isolation and there is no direct attempt at a global description of the data. Thus we could conclude that an association rule is a local pattern.

Consider now an association rule with a very high support, covering a large part of the data and representing some obvious knowledge that domain experts are already aware of: is this still a local pattern?

From the above considerations two questions arise:

1. *How much local is an association rule?*
2. *When does an association rule really represents some interesting knowledge?*

The two questions correspond to the first two features (listed above) that a structure must exhibit in order to be considered a local pattern: *locality* and *deviation*. Therefore in the following we will discuss the concepts of locality and deviation when applied to association rules and frequent patterns.

Locality (and Frequency) - In [14] Hand states: *"In order to define what is meant by local it is necessary to adopt a suitable distance measure. This choice will depend on the data and the application domain: in the cases of categorical variables, it might even require exact matches. It is also necessary to pick a threshold with which the measured distance is compared."*

When talking about frequent patterns is quite obvious to think about frequency as the measure of locality: a very frequent pattern is global (i.e. it covers a large part of the data), a not so frequent pattern is local (i.e. it covers only a small portion of the data). This agrees with the position of Morik who defined a local pattern as a description of *rare* events, which deviate from a global model and show an internal structure.

The need for mining patterns at very low support levels is confirmed also by the applications. In [22,23] association rules are extracted in a medical context. The authors state that most relevant rules with high quality (domain-dependent) metrics appear only at low frequency levels. In [10] association rules are mined with the aim of finding interesting associations between road number, weather and light conditions and serious or fatal accidents. The authors state that interesting (according to feedback from end-user) association rules were found only at very low support levels.

Therefore, a local pattern is a *"rare"* or *"not so frequent"* pattern. But at the same time it needs some support in order to distinguish by the mere random component. So a local pattern must be frequent but not too much.

These last considerations lead us to reformulate Hand's definition by the point of view of frequency:

$$data = very\_frequent\_patterns + rare\_patterns + random\_component$$

But how can reasonable support thresholds be defined? Hand states [14]: *"Some sort of compromise is needed. One way to find a suitable compromise would be to gradually expand the distance threshold defining local, so that the set of points identified as possible patterns gradually increases, stopping when the detected number of patterns seemed reasonable (a decision which depends on resources as well as on the phenomenon under investigation)."*

Deviation (and Constraints) - Extracting too many uninteresting frequent patterns, with large requirements both in terms of time and space, is an even harder problem when mining at very low support level. In fact, the pruning power of the frequency constraint decreases together with the minimum support threshold. When it becomes too low, the search space explodes and the computation becomes intractable.

This decrease in the pruning power of the frequency constraint, can be compensated by the pruning power of other constraints that the user could exploit to restrict the search for interesting patterns.

Example 1. Consider a student database at a university: rows corresponds to students, columns to courses, and a 1 entry (s, c) indicates that student s has taken course c. In other words, students are the transactions and courses taken are items in the transactions. In this context a pattern is a set of courses which satisfy some interestingness conditions. Frequent patterns are set of courses which appear together in the curricula of a number of students larger than a given minimum support threshold.

However the search for patterns can be guided by other interestingness constraints. Suppose that each course has some attributes, such has *semester*, *credits*, *prerequisites*, *difficulty_rate*. One could be interested in finding frequent set of courses c such that $c.semester = 1$ and $sum(c.credits) \geq 50$. These constraints can be pushed inside the frequent pattern algorithm, reducing the search space and thus enabling mining at low frequency levels.

Constraints are not only useful to prune the search space, thus reducing the computation. They have also a semantic value since the language of constraints is what the user exploits in order to define which are the *interesting* patterns.

The importance of constraints in the search for local patterns is confirmed by other definitions which emerged at the Dagstuhl seminar. According to Siebes *"local patterns are described by structural requirements, virtual attributes, and conditions on attribute values."* Similarly Boulicaut states that a local pattern is *"a sentence from a pattern language that is apriori interesting since it satisfies a given set of constraints and tells something about part of the data."*

As stated before, one important feature of local patterns, is deviation. Constraints can be our guide in the search for deviating patterns.

Example 2. Consider again our student/courses database. Suppose the average *difficulty_rate* of courses to be 0.35. We can search for set of courses c, (not so) frequently taken together, and such that $avg(c.difficulty\_rate) \geq 0.95$. If we find a solution to this pattern query, this is clearly an interesting deviating pattern.

The idea of the previous example might be generalized by the following naïve definition.

Definition 3 (Deviating Pattern). *Let* $\mathcal{I} = \{x_1, ..., x_n\}$ *be a set of items. Let* A_i *be a non-negative real-valued attribute associated to each item. Given a minimum deviation factor* $\delta > 1$; *an itemset* X *is said to be deviating on attribute* A_i *if:*

$$avg(X.A_i) \geq \delta \cdot avg(\mathcal{I}.A_i) \quad or \ if \quad avg(X.A_i) \leq \frac{avg(\mathcal{I}.A_i)}{\delta}$$

Since the value of $avg(\mathcal{I}.A_i)$ is just a constant, known before the mining phase, we can simply search for frequent itemsets which satisfy a given constraint defined on the average aggregate on some attribute of items. This is the kind of constraint which is usually studied in constrained frequent pattern mining [8,24,25]. Obviously the previous is just a naïve definition geared on Example 2: however, other similar measures of interestingness or deviation, might be defined w.r.t. the current application.

Local Patterns = (Not So) Frequent, Deviating Patterns - As stated above, in order to detect local interesting patterns, we can search for frequent itemsets at very low support threshold, which satisfy some deviation constraint. Recalling Hand's definition [14]:

$$data = background\_model + local\_patterns + random\_component$$

we can say that local patterns are:

(not so) frequent: a very low minimum support threshold is exploited to discard the random component;

deviating: some deviation constraint is exploited to discard the background model and to make local patterns emerge.

Summarizing, in our vision constraints play a twofold fundamental role in the search for local patterns: they enable mining at very low frequency levels, which is where local patterns are; and at the same time, they guide the search towards deviating or interesting patterns.

According to this vision, local patterns can be characterized as:

"(not so) frequent patterns which exhibit a deviating behavior."

3   Constraint Pushing Techniques

Constrained frequent pattern mining can be seen as a query optimization problem: given a mining query $\mathcal{Q}$ containing a set of constraints $\mathcal{C}$, provide an efficient evaluation strategy for $\mathcal{Q}$ which is sound and complete (i.e. it finds all and only itemsets in $Th(\mathcal{C}_{freq}) \cap Th(\mathcal{C})$). A naïve solution to such a problem is to first find all frequent patterns ($Th(\mathcal{C}_{freq})$) and then test them for constraints satisfaction. However more efficient solutions can be found by analyzing the property of constraints comprehensively, and exploiting such properties in order to push constraints in the frequent pattern computation. Following this methodology, some classes of constraints which exhibit nice properties have been individuated. In this Section, by reviewing all basic works on the constrained frequent itemsets mining problem, we recall a classification of constraints and their properties.

3.1   Anti-monotone and Succinct Constraints

A first work defining classes of constraints which exhibit nice properties is [21]. In that paper is introduced an Apriori-like algorithm, named CAP, which exploits two properties of constraints, namely *anti-monotonicity* and *succinctness*, in order to reduce the frequent itemsets computation. Four classes of constraints, each one with its own associated computational strategy, are defined:

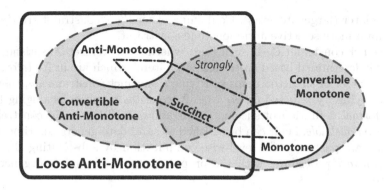

Fig. 1. Characterization of the classes of commonly used constraints.

1. Anti-monotone but not succinct constraints;
2. Anti-monotone and succinct constraints;
3. Succinct but not anti-monotone constraints;
4. Constraints that are neither.

Definition 4 (Anti-monotone constraint). *Given an itemset X, a constraint $\mathcal{C}_{AM}$ is anti-monotone if $\forall Y \subseteq X : \mathcal{C}_{AM}(X) \Rightarrow \mathcal{C}_{AM}(Y)$.*

The frequency constraint is the most known example of a $\mathcal{C}_{AM}$ constraint. This property, *the anti-monotonicity of frequency*, is used by the Apriori [2] algorithm with the following heuristic: if an itemset X does not satisfy $\mathcal{C}_{freq}$, then no superset of X can satisfy $\mathcal{C}_{freq}$, and hence they can be pruned. This pruning can affect a large part of the search space, since itemsets form a lattice. Therefore the Apriori algorithm (see Algorithm1) operates in a level-wise fashion moving bottom-up on the itemset lattice, from small to large itemsets. At each iteration k Apriori counts the support of *candidate* itemsets (i.e. itemsets which have all subsets frequent) of size k, which are denoted by C_k. Those ones which have a support greater than the minimum support threshold σ are frequent itemsets. From the set of frequent itemsets of size k (denoted by L_k) the set of candidates for the next iteration C_{k+1} is generated by the *generate_apriori* procedure.

Other $\mathcal{C}_{AM}$ constraints can easily be pushed deeply down into the frequent itemsets mining computation since they behave exactly as $\mathcal{C}_{freq}$: if they are not satisfiable at an early level (small itemsets), they have no hope of becoming

---

Algorithm 1 Apriori

Input: $\mathcal{D}, \sigma$
Output: $Th(\mathcal{C}_{freq[\mathcal{D},\sigma]})$
1: $C_1 \leftarrow \{\{i\} \mid i \in \mathcal{I}\}; \ k \leftarrow 1$
2: **while** $C_k \neq \emptyset$ **do**
3: $\quad L_k \leftarrow count(\mathcal{D}, C_k)$
4: $\quad C_{k+1} \leftarrow generate\_apriori(L_k)$
5: $\quad k++$
6: $Th(\mathcal{C}_{freq[\mathcal{D},\sigma]}) \leftarrow \bigcup_k L_k$

---

satisfiable later (larger itemsets). Conjoining other $\mathcal{C}_{AM}$ constraints to $\mathcal{C}_{freq}$ we just obtain a more selective anti-monotone constraint.

A succinct constraint $\mathcal{C}_S$ is such that, whether an itemset X satisfies it or not, can be determined based on the singleton items which are in X. Informally, given A_1, the set of singleton items satisfying a succinct constraint $\mathcal{C}_S$, then any set X satisfying $\mathcal{C}_S$ is based on A_1, i.e. X contains a subset belonging to A_1 (for the formal definition of succinct constraints see [21]). A $\mathcal{C}_S$ constraint is *pre-counting pushable*, i.e. it can be satisfied at candidate-generation time: these constraints are pushed in the level-wise computation by substituting the usual *generate_apriori* procedure, with the proper (w.r.t. $\mathcal{C}_S$) candidate generation procedure.

For instance, consider the constraint $\mathcal{C}_S \equiv min(X.price) \leq v$, which is a succinct but not anti-monotone constraint. Given $A_1 = \{i \in \mathcal{I} \mid i.price \leq v\}$, we have that $Th(\mathcal{C}_S) = \{X \in 2^{\mathcal{I}} \mid \exists i \in X : i \in A_1\}$. Therefore this constraint can be satisfied at candidate-generation time. This can be done using a special candidate generation procedure, which takes care of the kind of the given constraint, and produces only candidates which satisfy it.

Constraints that are both anti-monotone and succinct can be pushed completely in the level-wise computation before it starts (at pre-processing time). For instance, consider the constraint $min(X.price) \geq v$: if we start with the first set of candidates formed by all singleton items having price greater than v, during the computation we will generate only itemsets satisfying the given constraint.

Constraints that are neither succinct nor anti-monotone are pushed in the CAP [21] computation by inducing weaker constraints which are either anti-monotone and/or succinct.

3.2   Monotone Constraints

Monotone constraints work the opposite way of anti-monotone constraints.

Definition 5 (Monotone constraint). *Given an itemset X, a constraint $\mathcal{C}_M$ is monotone if:* $\forall Y \supseteq X : \mathcal{C}_M(X) \Rightarrow \mathcal{C}_M(Y)$.

Since the frequent itemset computation is geared on $\mathcal{C}_{freq}$, which is anti-monotone, $\mathcal{C}_M$ constraints have been considered more hard to be pushed in the computation and less effective in pruning the search space. In fact, many works [3,27,9,16] have studied the computational problem $Th(\mathcal{C}_{freq}) \cap Th(\mathcal{C}_M)$, proposing some smart exploration of its search space, but all facing the inherent difficulty of the computational problem: the $\mathcal{C}_{AM}$-$\mathcal{C}_M$ *tradeoff*.

The $\mathcal{C}_{AM}$-$\mathcal{C}_M$ Tradeoff - Such tradeoff can be described as follows. Suppose that an itemset has been removed from the search space because it does not satisfy a monotone constraint. This pruning avoids checking support for this itemset, but on the other hand, if we check its support and find it smaller than the frequency threshold, we may prune away all the supersets of this itemset.

In other words, by monotone pruning we risk to lose anti-monotone pruning opportunities given by the pruned itemset.

The tradeoff is clear: pushing monotone constraint can save frequency tests, however the results of these tests could have lead to more effective anti-monotone pruning.

The ExAnte Property and the $\mathcal{C}_{AM}$-$\mathcal{C}_M$ Synergy - In [5] a completely new approach to exploit monotone constraints by means of data-reduction is introduced. The *ExAnte Property* [5] is obtained by shifting attention from the pattern search space to the input data. Indeed, the $\mathcal{C}_{AM}$-$\mathcal{C}_M$ *tradeoff* exists only if we focus exclusively on the search space of the problem, while if exploited properly, monotone constraints can reduce dramatically the data in input, in turn strengthening the anti-monotonicity pruning power. With data reduction techniques we exploit the effectiveness of a $\mathcal{C}_{AM}$-$\mathcal{C}_M$ *synergy*.

The ExAnte property states that a transaction which does not satisfy the given monotone constraint can be deleted from the input database since it will never contribute to the support of any itemset satisfying the constraint.

Proposition 1 (ExAnte property [5]). *Given a transaction database $\mathcal{D}$ and a conjunction of monotone constraints $\mathcal{C}_M$, we define the μ-reduction of $\mathcal{D}$ as the dataset resulting from pruning the transactions that do not satisfy $\mathcal{C}_M$: $\mu_{\mathcal{C}_M}(\mathcal{D}) = \{t \in \mathcal{D} \mid t \in Th(\mathcal{C}_M)\}$.*
It holds that this data reduction does not affect the support of solution itemsets:

$$\forall X \in Th(\mathcal{C}_M) : supp_{\mathcal{D}}(X) = supp_{\mu_{\mathcal{C}_M}(\mathcal{D})}(X).$$

A major consequence of reducing the input database in this way is that it implicitly reduces the support of a large amount of itemsets that do not satisfy $\mathcal{C}_M$ as well, resulting in a reduced number of candidate itemsets generated during the mining algorithm. Even a small reduction in the database can cause a huge cut in the search space, because all supersets of infrequent itemsets are pruned from the search space as well. In other words, monotonicity-based data-reduction of transactions strengthens the anti-monotonicity-based pruning of the search space.

This is not the whole story, in fact, infrequent singleton items can not only be removed from the search space together with all their supersets, for the same anti-monotonicity property they also can be deleted from all transactions in the input database (this anti-monotonicity-based data-reduction is named α-*reduction*). Removing items from transactions offers another positive effect: reducing the size of a transaction which satisfies $\mathcal{C}_M$ can make the transaction violate it. Therefore a growing number of transactions which do not satisfy $\mathcal{C}_M$ can be found. Obviously, we are inside a loop where two different kinds of pruning (α and μ) cooperate to reduce the search space and the input dataset, strengthening each other step by step until no more pruning is possible (a fix-point has been reached). This is the key idea of the ExAnte pre-processing method [5]. In

item	price
a	5
b	8
c	14
d	30
e	20
f	15
g	6
h	12

(a)

item	1_{st}	2_{nd}	3_{rd}
a	3	†	†
b	7	4	4
c	5	5	4
d	7	5	4
e	3	†	†
f	3	†	†
g	5	3	†
h	2	†	†

(b)

tID	Itemset	Total price
1	b,c,d,g	58
2	a,b,d,e	63
3	b,c,d,g,h	70
4	a,e,g	31
5	c,d,f,g	65
6	a,b,c,d,e	77
7	a,b,d,f,g,h	76
8	b,c,d	52
9	b,e,f,g	49

(c)

tID	Itemset	Total price
1	b,c,d,g	58
2	b,d	38
3	b,c,d,g	58
5	c,d,g	50
6	b,c,d	52
7	b,d,g	44
8	b,c,d	52
9	b,g	14

(d)

tID	Itemset	Total price
1	b,c,d,g	58
3	b,c,d,g	58
5	c,d,g	50
6	b,c,d	52
8	b,c,d	52

(e)

tID	Itemset	Total price
1	b,c,d	52
3	b,c,d	52
6	b,c,d	52
8	b,c,d	52

(f)

Table 1. Example: the price table (a), items and their supports iteration by iteration (b), the initial transaction database (c), two intermediate status of the database (d) and (e), and the final preprocessed database (f).

the end, the reduced dataset resulting from this fix-point computation is usually much smaller than the initial dataset, and it can feed any frequent itemset mining algorithm for a much smaller (but complete) computation.

Suppose we have the transaction database in Table 1 (c) and the price database in Table 1 (a). Suppose that we want to compute frequent itemsets ($min\_supp = 4$) with a sum of prices ≥ 45. During the first iteration the total price of each transaction is checked to avoid using transactions which do not satisfy the monotone constraint. All transaction with a sum of prices ≥ 45 are used to count the support for the singleton items. Only the fourth transaction is discarded. At the end of the count we find items a, e, f and h to be infrequent. Note that, if the fourth transaction had not been discarded, items a and e would have been counted as frequent. At this point we perform an α-reduction of the database: this means removing a, e, f and h from all transactions obtaining reduced database in Table 1 (d). After the α-reduction we have more opportunities to μ-reduce the database. In fact transaction 2, which at the beginning has a total price of 63, now has its total price reduced to 38 due to the pruning of a and e. This transaction can now be pruned away. The same reasoning holds for transactions number 7 and 9. We obtain database in Table 1 (e). At this point ExAnte counts once again the support of alive items within the reduced database. The item g which initially has got a support of 5 now has become infrequent (see Table 1 (b) for items support iteration by iteration). We can α-reduce again the database, and then μ-reduce it. After the two reductions transaction number 5 does not satisfy anymore the monotone constraint and it is pruned away (see database in Table 1 (f)). ExAnte counts again the support of items within the reduced database but no more items are found to have turned infrequent. The fix-point has been reached at the third iteration: the database has been reduced from 9 transactions to 4 transactions (number 1,3,6 and 8), and interesting itemsets have shrunk from 8 to 3 (b, c and d). The final database contains only the unique solution to problem which is the 3-itemset $\{b, c, d\}$ with support 4 and sum of prices 52. Note that on this toy-example we even do not need to run a mining algorithm!

This simple yet very effective idea has been generalized from pre-processing to effective mining in two main directions: in an Apriori-like breadth-first computation in *ExAMiner* [4], and in a FP-growth [13] based depth-first computation in *FP-Bonsai* [6].

ExAMiner - The recently introduced algorithm ExAMiner [4], generalizes the ExAnte idea to reduce the problem dimensions at all levels of a level-wise Apriori-like computation. In this way, the $\mathcal{C}_{AM}$-$\mathcal{C}_M$ *synergy* is effectively exploited at each iteration of the mining algorithm, and not only at pre-processing as done by ExAnte, resulting in significant performance improvements.

The idea is to generalize ExAnte's α-reduction from singletons level to the generic level k. This generalization results in the following set of data reduction techniques, which are based on the anti-monotonicity of $\mathcal{C}_{freq}$ (see [4] for the proof of correctness).

$\mathcal{G}_k(i)$: an item which is not subset of at least k frequent k-itemsets can be pruned away from all transactions in $\mathcal{D}$.

$\mathcal{T}_k(t)$: a transaction which is not superset of at least $k + 1$ frequent k-itemsets can be removed from $\mathcal{D}$.

$\mathcal{L}_k(i)$: given an item i and a transaction t, if the number of frequent k-itemsets which are superset of i and subset of t is less than k, then i can be pruned away from transaction t.

In ExAMiner [4] these data reductions are coupled with the μ-reduction for $\mathcal{C}_M$ constraints as described in Proposition 1.

Essentially ExAMiner is an Apriori-like algorithm, which at each iteration $k - 1$ produces a reduced dataset $\mathcal{D}_k$ to be used at the subsequent iteration k. Each transaction in $\mathcal{D}_k$, before participating to the support count of candidate itemsets, is reduced as much as possible by means of $\mathcal{C}_{freq}$-based data reduction, and only if it survives to this phase, it is effectively used in the counting phase. Each transaction which arrives to the counting phase, is then tested against the $\mathcal{C}_M$ (μ-reduction) , and reduced again as much as possible, and only if it survives to this second set of reductions, it is written to the transaction database for the next iteration $\mathcal{D}_{k+1}$. The procedure we have just described, is named *count&reduce* (see Algorithm 2), and substitutes the usual support counting procedure of the Apriori algorithm. In Algorithm 2 in order to implement the data-reduction $\mathcal{G}_k(i)$ we use an array of integers V_k (of the size of *Items*), which records for each item the number of frequent k-itemsets in which it appears. This information is then exploited during the subsequent iteration $k+1$ for the global pruning of items from all transaction in $\mathcal{D}_{k+1}$ (lines 3 and 4 of the pseudo-code). On the contrary, data reductions $\mathcal{T}_k(t)$ and $\mathcal{L}_k(i)$ are put into effect during the

---

Algorithm 2 *count&reduce*

Input: $\mathcal{D}_k, \sigma, \mathcal{C}_M, C_k, V_{k-1}$

1: **forall** $i \in \mathcal{I}$ **do** $V_k[i] \leftarrow 0$
2: **forall** tuples t in $\mathcal{D}_k$ **do**
3:     **forall** $i \in t$ **do if** $V_{k-1}[i] < k - 1$
4:       **then** $t \leftarrow t \setminus i$
5:       **else** $i.count \leftarrow 0$
6:     **if** $|t| \geq k$ **and** $\mathcal{C}_M(t)$ **then**
7:       **forall** $X \in C_k, X \subseteq t$ **do**
8:         $X.count$++; $t.count$++
9:         **forall** $i \in X$ **do** $i.count$++
10:       **if** $X.count = \sigma$ **then**
11:         $L_k \leftarrow L_k \cup \{X\}$
12:         **forall** $i \in X$ **do** $V_k[i]$++
13:       **if** $|t| \geq k + 1$ **and** $t.count \geq k + 1$ **then**
14:         **forall** $i \in t$ **if** $i.count < k$
15:           **then** $t \leftarrow t \setminus i$
16:           **if** $|t| \geq k + 1$ **and** $\mathcal{C}_M(t)$ **then**
17:             **write** t in $\mathcal{D}_{k+1}$

---

same iteration in which the information is collected. Unfortunately, they require information (the frequent itemsets of cardinality k) that is available only at the end of the actual counting (when all transactions have been used). However, since the set of frequent k-itemsets is a subset of the set of candidates C_k, we can use such data reductions in a relaxed version: we just check the number of candidate itemsets X which are subset of t ($t.count$ in the pseudo-code, lines 10 and 18) and which are superset of i ($i.count$ in the pseudo-code, lines 9 and 14).

FP-bonsai - In [6] it is shown how the $\mathcal{C}_{AM}$-$\mathcal{C}_M$ synergy can be exploited within the well known FP-growth algorithm [13]. Thanks to the recursive projecting approach of FP-growth, the ExAnte data-reduction is pervasive all over the computation. All the FP-trees built recursively during the FP-growth computation can be pruned extensively by using the ExAnte property (Proposition 1), obtaining a computation with a smaller number of smaller trees. Such tiny FP-trees, obtained by growing and pruning, are named *FP-bonsai*. The resulting method overcomes on one hand the main drawback of FP-growth, which is its memory requirements, and on the other hand, the main drawback of ExAMiner which is the I/O cost of iteratively rewriting the reduced datasets to disk.

3.3 Convertible Constraints

In [24,25] a new class of tough constraints is introduced and it is shown how such constraints can be pushed within a FP-growth [13] computation. Consider the constraint defined on the *average* aggregate (e.g. $avg(X.price) \leq v$): it is quite straightforward to show that it is not anti-monotone, nor monotone, nor succinct. Subsets (or supersets) of a valid itemset could well be invalid and vice versa. For this reason, in [24,25] the authors state that within the level-wise framework, no direct pruning based on such constraints can be made. But, if we arrange the items in *price-descending-order* we can observe an interesting property: the average of an itemset is no more than the average of its prefix itemset, according to this order. The FP-growth approach to frequent itemset mining, is based on the concept of prefix-itemsets, therefore its quite easy to integrate *convertible* constraints in such an algorithmic framework.

Definition 6 (Prefix itemset). *Given a total order $\mathcal{R}$ over $\mathcal{I}$, an itemset $X' = i_1 i_2 ... i_l$ is called a prefix of itemset $X = i_1 i_2 ... i_m$ w.r.t. $\mathcal{R}$, where ($l \leq m$) and items in both itemsets are listed according to order $\mathcal{R}$.*

Definition 7 (Convertible constraints). *A constraint $\mathcal{C}_{CAM}$ is convertible anti-monotone provided there is an order $\mathcal{R}$ on items such that whenever an itemset X satisfies $\mathcal{C}_{CAM}$, so does any prefix of X. A constraint $\mathcal{C}_{CM}$ is convertible monotone provided there is an order $\mathcal{R}$ on items such that whenever an itemset X violates $\mathcal{C}_{CM}$, so does any prefix of X.*

In order to be convertible, a constraint must be defined over a *Prefix Increasing (resp. Decreasing) Function*, i.e. a function $f : 2^{\mathcal{I}} \to \mathbb{R}$ such that for every itemset S and item a, if $\forall x \in S, x \mathcal{R} a$ then $f(S) \leq$ (resp. $\geq$) $f(S \cup \{a\})$. Let f be a prefix increasing (resp. decreasing) function w.r.t. a given order $\mathcal{R}$. Then $f(X) \geq v$ is a convertible monotone (resp. anti-monotone) constraint, while $f(X) \leq v$ is a convertible anti-monotone (resp. monotone) constraint.

Proposition 2. *Any anti-monotone (resp. monotone) constraint is trivially convertible anti-monotone (resp. convertible monotone): just pick any order on items.*

Example 3 (avg constraint is convertible). Let $\mathcal{R}$ be the value-descending order. Given an itemset $X = i_1 i_2 ... i_l$ satisfying the constraint $avg(X) \geq v$, where items in X are listed in order $\mathcal{R}$. For each prefix $X' = i_1 i_2 ... i_k$ of X ($1 \leq k \leq l$), since $i_k \geq i_{k+1} \geq ... \geq i_l$ we have that $avg(X') \geq avg(X) \geq v$, thus also X' satisfies the constraint. This implies that $avg(X) \geq v$ is a $\mathcal{C}_{CAM}$ constraint. Similarly it can be shown that $avg(X) \leq v$ is $\mathcal{C}_{CM}$ w.r.t. the same order.

Interestingly, if the order $\mathcal{R}^{-1}$ (i.e. the reversed order of $\mathcal{R}$) is used, the constraint $avg(S) \geq v$ can be shown convertible monotone, and $avg(S) \leq v$ convertible anti-monotone. Constraints which exhibit this interesting property of being convertible in both a monotone or an anti-monotone constraints, are called *strongly convertible*.

Definition 8 (Strongly convertible constraints). *A constraint $\mathcal{C}$ is strongly convertible provided there is an order $\mathcal{R}$ over the set of items such that $\mathcal{C}$ is convertible anti-monotone w.r.t. $\mathcal{R}$ and convertible monotone w.r.t. $\mathcal{R}^{-1}$.*

Constraint	Anti-monotone	Monotone	Succinct	Convertible	C^l_{LAM}
$min(S.A) \geq v$	yes	no	yes	strongly	$l = 1$
$min(S.A) \leq v$	no	yes	yes	strongly	$l = 1$
$max(S.A) \geq v$	no	yes	yes	strongly	$l = 1$
$max(S.A) \leq v$	yes	no	yes	strongly	$l = 1$
$count(S) \leq v$	yes	no	weakly	$\mathcal{A}$	$l = 1$
$count(S) \geq v$	no	yes	weakly	$\mathcal{M}$	$l = v$
$sum(S.A) \leq v \ (\forall i \in S, i.A \geq 0)$	yes	no	no	$\mathcal{A}$	$l = 1$
$sum(S.A) \geq v \ (\forall i \in S, i.A \geq 0)$	no	yes	no	$\mathcal{M}$	no
$sum(S.A) \leq v \ (v \geq 0, \forall i \in S, i.A\theta 0)$	no	no	no	$\mathcal{A}$	$l = 1$
$sum(S.A) \geq v \ (v \geq 0, \forall i \in S, i.A\theta 0)$	no	no	no	$\mathcal{M}$	no
$sum(S.A) \leq v \ (v \leq 0, \forall i \in S, i.A\theta 0)$	no	no	no	$\mathcal{M}$	no
$sum(S.A) \geq v \ (v \leq 0, \forall i \in S, i.A\theta 0)$	no	no	no	$\mathcal{A}$	$l = 1$
$range(S.A) \leq v$	yes	no	no	strongly	$l = 1$
$range(S.A) \geq v$	no	yes	no	strongly	$l = 2$
$avg(S.A)\theta v$	no	no	no	strongly	$l = 1$
$median(S.A)\theta v$	no	no	no	strongly	$l = 1$
$var(S.A) \geq v$	no	no	no	no	$l = 2$
$var(S.A) \leq v$	no	no	no	no	$l = 1$
$std(S.A) \geq v$	no	no	no	no	$l = 2$
$std(S.A) \leq v$	no	no	no	no	$l = 1$
$var_{N-1}(S.A)\theta v$	no	no	no	no	$l = 2$
$md(S.A) \geq v$	no	no	no	no	$l = 2$
$md(S.A) \leq v$	no	no	no	no	$l = 1$

Table 2. Classification of commonly used constraints (where $\theta \in \{\geq, \leq\}$, k denotes itemsets cardinality).

In [24,25], two FP-growth based algorithms are introduced: $\mathcal{FIC}^{\mathcal{A}}$ to mine $Th(\mathcal{C}_{freq}) \cap Th(\mathcal{C}_{CAM})$, and $\mathcal{FIC}^{\mathcal{M}}$ to mine $Th(\mathcal{C}_{freq}) \cap Th(\mathcal{C}_{CM})$.

A major limitation of any FP-growth based algorithm is that the initial database (internally compressed in the prefix-tree structure) and all intermediate projected databases must fit into main memory. If this requirement cannot be met, these approaches can simply not be applied anymore. This problem is even harder with $\mathcal{FIC}^{\mathcal{A}}$ and $\mathcal{FIC}^{\mathcal{M}}$: in fact, using an order on items different from the frequency-based one, makes the prefix-tree lose its compressing power. Thus we have to manage much greater data structures, requiring a lot more main memory which might not be available. This fact is confirmed by the experimental analysis reported in [8]: sometimes $\mathcal{FIC}^{\mathcal{A}}$ is slower than FP-growth, meaning that having constraints brings no benefit to the computation.

Another important drawback of this approach is that it is not possible to take full advantage of a conjunction of different constraints, since each constraint in the conjunction could require a different ordering of items.

3.4   Loose Anti-monotone Constraints

In [8] a new class of tougher constraints, which is a proper superclass of convertible anti-monotone, is introduced.

Example 4 (var constraint is not convertible). Calculating the variance is an important task of many statistical analysis: it is a measure of how spread out a distribution is.

The variance of a set of number X is defined as: $var(X) = \frac{\sum_{i \in X}(i - avg(X))^2}{|X|}$. A constraint based on var is not convertible. Otherwise there is an order $\mathcal{R}$ of items such that $var(X)$ is a prefix increasing (or decreasing) function. Consider a small dataset with only four items $\mathcal{I} = \{A, B, C, D\}$ with associated prices $P = \{10, 11, 19, 20\}$. The lexicographic order $\mathcal{R}_1 = \{ABCD\}$ is such that $var(A) \leq var(AB) \leq var(ABC) \leq var(ABCD)$, and it is easy to see that we have only other three orders with the same property: $\mathcal{R}_2 = \{BACD\}, \mathcal{R}_3 = \{DCBA\}, \mathcal{R}_4 = \{CDBA\}$. But, for $\mathcal{R}_1$, we have that $var(BC) \nleq var(BCD)$, which means that var is not a prefix increasing function w.r.t. $\mathcal{R}_1$. Moreover, since the same holds for $\mathcal{R}_2, \mathcal{R}_3, \mathcal{R}_4$, we can assert that there is no order $\mathcal{R}$ such that var is prefix increasing. An analogous reasoning can be used to show that it neither exists an order which makes var a prefix decreasing function.

Following a similar reasoning we can show that other interesting constraints, such as for instance those ones based on *standard deviation (std)* or *unbiased variance estimator (var$_{N-1}$)* or *mean deviation (md)*, are not convertible as well. The above example shows that such interesting constraints cannot be exploited within a prefix pattern framework. Luckily, as we show in the following, all these constraints share a nice property named *"Loose Anti-monotonicity"*. Recall that an anti-monotone constraint is such that, if satisfied by an itemset then it is satisfied by *all* its subsets. A loose anti-monotone constraint is such that, if it is satisfied by an itemset of cardinality k then it is satisfied by *at least one* of its subsets of cardinality $k - 1$.

Definition 9 (Loose Anti-monotone constraint). *Given an itemset X, a constraint is loose anti-monotone from size $l > 0$ (denoted $\mathcal{C}_{LAM}^l$) if:*

$$(|X| > l \land \mathcal{C}_{LAM}(X)) \Rightarrow \exists i \in X : \mathcal{C}_{LAM}(X \setminus \{i\})$$

The next proposition and the subsequent example state that the class of $\mathcal{C}_{LAM}^l$ constraints is a proper superclass of $\mathcal{C}_{CAM}$ (convertible anti-monotone constraints).

Proposition 3. *Any convertible anti-monotone constraint is trivially loose anti-monotone: if a k-itemset satisfies the constraint so does its $(k-1)$-prefix itemset.*

Example 5. We show that the constraint $var(X.A) \leq v$ is a $\mathcal{C}_{LAM}^1$ constraint. Given an itemset X, if it satisfies the constraint so trivially does $X \setminus \{i\}$, where i is the element of X which has associated a value of A which is the most far away from $avg(X.A)$. In fact, we have that $var(\{X \setminus \{i\}\}.A) \leq var(X.A) \leq v$, until $|X| > 1$, i.e. until $var(X \setminus \{i\})$ is defined. Taking the element of X which has associated a value of A which is the closest to $avg(X.A)$ we can show that $var(X.A) \geq v$ is a $\mathcal{C}_{LAM}^2$ constraint. In this case we have that the constraint is *loose anti-monotone from size 2* because the variance of a singleton item is zero. Since the standard deviation std is the square root of the variance, it is straightforward to see that $std(X.A) \leq v$ is $\mathcal{C}_{LAM}^1$, and $std(X.A) \geq v$ is $\mathcal{C}_{LAM}^2$.

The mean deviation is defined as: $md(X) = (\sum_{i \in X} |i - avg(X)|) \, / \, |X|$. Once again, we have that $md(X.A) \leq v$ is $\mathcal{C}^1_{LAM}$, and $md(X.A) \geq v$ is $\mathcal{C}^2_{LAM}$. It is easy to prove that also constraints defined on the unbiased variance estimator, $var_{N-1} = (\sum_{i \in X} (i - avg(X))^2) \, / \, (|X| - 1)$ are loose anti-monotone. In particular, they are $\mathcal{C}^2_{LAM}$ since they are not defined for singleton items.

In Table 2 and Figure 1 we provide the state-of-art classification of commonly used constraints.

The next Proposition indicates how a $\mathcal{C}^l_{LAM}$ constraint can be exploited in a level-wise Apriori-like computation by means of data-reduction. It states that if at a certain iteration $k > l$ a transaction is not superset of at least one frequent k-itemset which satisfy the $\mathcal{C}^l_{LAM}$ constraint (a solution), then the transaction can be deleted from the database.

Proposition 4. *Given a transaction database* $\mathcal{D}$, *a minimum support threshold* σ, *and a* $\mathcal{C}^l_{LAM}$ *constraint, at the iteration* $k > l$ *of the level-wise computation, a transaction* $t \in \mathcal{D}$ *such that:*

$$\nexists X \subseteq t, |X| = k, X \in Th(\mathcal{C}_{freq[\mathcal{D},\sigma]}) \cap Th(\mathcal{C}^l_{LAM})$$

can be pruned away from $\mathcal{D}$, *since it will never be superset of any solution itemsets of cardinality* $> k$.

As in ExAMiner [4] the anti-monotonicity based data reductions are coupled with the μ-reduction for $\mathcal{C}_M$ constraints, similarly we can exploit the above Proposition for $\mathcal{C}_{LAM}$ constraints, by embedding such loose anti-monotonicity based data reduction with-in the *count&reduce* procedure (see Algorithm 2. As usual, the more data-reduction techniques the better: we can exploit them all together, and they strengthen each other; i.e. and the total benefit is always greater than the sum of the individual benefits.

4   On Going Work: A Pattern Discovery System

As discussed before, one of the most important drawback of the FP-growth based approach is that it is not possible to take full advantage of a conjunction of different constraints, since each constraint in the conjunction could require a different ordering of items. On the contrary, in the data-reduction based approach we can fully exploit different kind of constraints: the more constraints we have the stronger is the data-reduction effect. In particular:

- Anti-monotone ($\mathcal{C}_{AM}$) constraints are exploited to prune the level-wise exploration of the search space together with the frequency constraint ($\mathcal{C}_{freq}$) as described in [21] and in Section 3.1;
- Succinct ($\mathcal{C}_S$) constraints are exploited at candidate generation time as done in [21] and in Section 3.1 ;
- Monotone ($\mathcal{C}_M$) constraints are exploited by means of data reduction as done in [5,4] and in Section 3.2;

– Convertible anti-monotone ($\mathcal{C}_{CAM}$) and Loose anti-monotone ($\mathcal{C}_{LAM}$) constraints are exploited by means of data reduction as described in [8] and in Section 3.4.

Example 6. The constraint $range(S.A) \geq v \equiv max(S.A) - min(S.A) \geq v$, is both $\mathcal{C}_M$ and $\mathcal{C}^2_{LAM}$. Thus, when we mine frequent itemsets which satisfy such constraint we can exploit the benefit of having together, in the same $count\&reduce^{\mathcal{LAM}}$ procedure, the $\mathcal{C}_{freq}$-based data reductions, μ-reduction, and reduction based on $\mathcal{C}_{LAM}$. Consider now the constraint $max(S.A) \geq v$. This constraint is $\mathcal{C}_M$, $\mathcal{C}_S$ and $\mathcal{C}^1_{LAM}$. This means that we can exploit all these properties by using it as a succinct constraint at candidate generation time as done in [21], and using it as a monotone constraint and as a loose anti-monotone constraint by means of data-reduction at counting time.

At Pisa KDD Laboratory we are currently developing such unified computational framework (within the P^3D project[1]) which will be the efficient computational engine of PATTERNIST, a pattern discovery system equipped with a GUI, a PDQL (pattern discovery query language) and visualization tools for the extracted patterns.

References

1. R. Agrawal, T. Imielinski, and A. N. Swami. Mining association rules between sets of items in large databases. In *Proceedings of the 1993 ACM International Conference on Management of Data (SIGMOD'93)*, 1993.
2. R. Agrawal and R. Srikant. Fast Algorithms for Mining Association Rules in Large Databases. In *Proceedings of the Twentieth International Conference on Very Large Databases (VLDB'94)*, 1994.
3. F. Bonchi, F. Giannotti, A. Mazzanti, and D. Pedreschi. Adaptive Constraint Pushing in frequent pattern mining. In *Proceedings of the 7th European Conference on Principles and Practice of Knowledge Discovery in Databases (PKDD'03)*, 2003.
4. F. Bonchi, F. Giannotti, A. Mazzanti, and D. Pedreschi. ExAMiner: Optimized level-wise frequent pattern mining with monotone constraints. In *Proceedings of the Third IEEE International Conference on Data Mining (ICDM'03)*, 2003.
5. F. Bonchi, F. Giannotti, A. Mazzanti, and D. Pedreschi. ExAnte: Anticipated data reduction in constrained pattern mining. In *Proceedings of the 7th European Conference on Principles and Practice of Knowledge Discovery in Databases (PKDD'03)*, 2003.
6. F. Bonchi and B. Goethals. FP-Bonsai: the art of growing and pruning small fp-trees. In *Proceedings of the Eighth Pacific-Asia Conference on Knowledge Discovery and Data Mining (PAKDD'04)*, Sydney, Australia, 2004.
7. F. Bonchi and C. Lucchese. On closed constrained frequent pattern mining. In *Proceedings of the Fourth IEEE International Conference on Data Mining (ICDM'04)*, 2004.
8. F. Bonchi and C. Lucchese. Pushing tougher constraints in frequent pattern mining. Technical Report 2004-TR-63, ISTI-C.N.R., 2004. (Submitted to SDM'05).

---

[1] http://www-kdd.isti.cnr.it/p3d/index.html

9. C. Bucila, J. Gehrke, D. Kifer, and W. White. DualMiner: A dual-pruning algorithm for itemsets with constraints. In *Proceedings of the 8th ACM International Conference on Knowledge Discovery and Data Mining (SIGKDD'02)*, 2002.

10. P. Flach et al. On the road to knowledge: mining 21 years of uk traffic accident reports. In *Data Mining and Decision Support: Aspects of Integration and Collaboration*, pages 143–155. Kluwer Academic Publishers, January 2003.

11. G. Grahne, L. Lakshmanan, and X. Wang. Efficient mining of constrained correlated sets. In *16th IEEE International Conference on Data Engineering (ICDE'00)*, 2000.

12. J. Han, L. V. S. Lakshmanan, and R. T. Ng. Constraint-based, multidimensional data mining. *Computer*, 32(8):46–50, 1999.

13. J. Han, J. Pei, and Y. Yin. Mining frequent patterns without candidate generation. In *Proceedings of the 2000 ACM International Conference on Management of Data (SIGMOD'00)*, 2000.

14. D. Hand. Pattern detection and discovery. In *Proceedings of the ESF Exploratory Workshop on Pattern Detection and Discovery in Data Mining*, volume 2447 of *Lecture Notes in Computer Science*, 2002.

15. D. J. Hand, N. M. Adams, and R. J. Bolton, editors. *Pattern Detection and Discovery, ESF Exploratory Workshop, London, UK, September 16-19, 2002, Proceedings*, volume 2447 of *Lecture Notes in Computer Science*, 2002.

16. B. Jeudy and J.-F. Boulicaut. Optimization of association rule mining queries. *Intelligent Data Analysis Journal*, 6(4):341–357, 2002.

17. L. V. S. Lakshmanan, R. T. Ng, J. Han, and A. Pang. Optimization of constrained frequent set queries with 2-variable constraints. In *Proceedings of the ACM International Conference on Management of Data (SIGMOD'99)*, 1999.

18. W. Li, J. Han, and J. Pei. CMAR: Accurate and efficient classification based on multiple class-association rules. In *In Proceedings of the 2001 IEEE International Conference on Data Mining (ICDM'01)*, 2001.

19. B. Liu, W. Hsu, and Y. Ma. Integrating classification and association rule mining. In *4th Int. Conf. Knowledge Discovery and Data Mining (KDD'98)*, pages 80–86, New York, 1998.

20. H. Mannila. Local and global methods in data mining: Basic techniques and open problems. In *Automata, Languages and Programming, 29th International Colloquium, ICALP 2002, Malaga, Spain, July 8-13, 2002, Proceedings*, volume 2380 of *Lecture Notes in Computer Science*, 2002.

21. R. T. Ng, L. V. S. Lakshmanan, J. Han, and A. Pang. Exploratory mining and pruning optimizations of constrained associations rules. In *Proceedings of the ACM International Conference on Management of Data (SIGMOD'98)*, 1998.

22. C. Ordonez, L. de Braal, and C. A. Santana. Discovering interesting association rules in medical data. In *ACM SIGMOD Workshop on Research Issues in Data Mining and Knowledge Discovery (DMKD'00)*, 2000.

23. C. Ordonez, E. Omiecinski, L. de Braal, C. A. Santana, N. Ezquerra, J. A. Taboada, D. Cooke, E. Krawczynska, and E. V. Garcia. Mining constrained association rules to predict heart disease. In *Proceedings of the First IEEE International Conference on Data Mining (ICDM'01)*, 2001.

24. J. Pei and J. Han. Can we push more constraints into frequent pattern mining? In *Proceedings of the 6th ACM International Conference on Knowledge Discovery and Data Mining (SIGKDD'00)*, 2000.

25. J. Pei, J. Han, and L. V. S. Lakshmanan. Mining frequent item sets with convertible constraints. In *17th IEEE International Conference on Data Engineering (ICDE'01)*, 2001.

26. J. Pei, X. Zhang, M. Cho, H. Wang, and P. Yu. Maple: A fast algorithm for maximal pattern-based clustering. In *Proceedings of the Third IEEE International Conference on Data Mining (ICDM'03)*, 2003.
27. L. D. Raedt and S. Kramer. The levelwise version space algorithm and its application to molecular fragment finding. In *Proceedings of the Seventeenth International Joint Conference on Artificial Intelligence, (IJCAI'01)*, 2001.
28. R. Srikant, Q. Vu, and R. Agrawal. Mining association rules with item constraints. In *Proceedings of the 3rd ACM International Conference on Knowledge Discovery and Data Mining, (SIGKDD'97)*, 1997.
29. M. L. Yiu and N. Mamoulis. Frequent-pattern based iterative projected clustering. In *Proceedings of the Third IEEE International Conference on Data Mining (ICDM'03)*, 2003.

From Local to Global Patterns:
Evaluation Issues in Rule Learning Algorithms

Johannes Fürnkranz

TU Darmstadt, Knowledge Engineering Group
Hochschulstraße 10, D-64289 Darmstadt, Germany
fuernkranz@informatik.tu-darmstadt.de

Abstract. Separate-and-conquer or covering rule learning algorithms may be viewed as a technique for using local pattern discovery for generating a global theory. Local patterns are learned one at a time, and each pattern is evaluated in a local context, with respect to the number of positive and negative examples that it covers. Global context is provided by removing the examples that are covered by previous patterns before learning a new rule. In this paper, we discuss several research issues that arise in this context. We start with a brief discussion of covering algorithms, their problems, and review a few suggestions for resolving them. We then discuss the suitability of a well-known family of evaluation metrics, and analyze how they trade off coverage and precision of a rule. Our conclusion is that in many applications, coverage is only needed for establishing statistical significance, and that the rule discovery process should focus on optimizing precision. As an alternative to coverage-based overfitting avoidance, we then investigate the feasibility of meta-learning a predictor for the true precision of a rule, based on its coverage on the training set. The results confirm that this is a valid approach, but also point at some shortcomings that need to be addressed in future work.

1 Introduction

Numerous evaluation heuristics have been proposed for evaluating rules in the context of classification rule learning, subgroup discovery and association rule discovery (Fürnkranz, 1999; Lavrač et al., 1999; Tan et al., 2002). Nevertheless, the issue is not yet well understood. The similarities and differences between the different measures are not explored in sufficient depth, and it is often also not clear, what properties we want an evaluation measure to have. Our research aims at increasing our understanding of these issues through theoretical analysis as well as empirical evaluation.

In this paper, we first discuss the relation between classification rule discovery using the separate-and-conquer or covering strategy and the local pattern discovery task (Section 2). Following up on (Fürnkranz and Flach, 2005), our main tool for analysis will be visualization in coverage space (Section 3). In the following (Section 4), we will analyze a family of well-known rule evaluation measures that have been proposed for subgroup discovery (Klösgen, 1992; Wrobel, 1997),

K. Morik et al. (Eds.): Local Pattern Detection, LNAI 3539, pp. 20–38, 2005.

```
function COVERING(Examples)

# initialize the classifier
GlobalClassifier ← ∅

# loop until all examples are covered
while Examples ≠ ∅

    # find the best local pattern
    LocalPattern ← FINDBESTLOCALPATTERN(Examples)

    # add the local pattern to the classifier
    GlobalClassifier ← GlobalClassifier ∪ LocalPattern

    # remove the covered examples
    Examples ← Examples \ COVERED(LocalPattern, Examples)

return GlobalClassifier
```

Fig. 1. The Covering Algorithm

that all have in common that they trade off coverage and precision of the rule, but differ in the weight that they allot to each component. We will argue that, with increasing coverage, the relative importance of coverage becomes negligible. One of the main reasons for including coverage is the problem of overfitting, to which precision is particularly susceptible. In the final part of the paper (Section 5), we address this problem in a novel way, namely by using training set statistics of a rule for predicting its test performance.

2   From Local to Global Patterns

The covering or separate-and-conquer strategy for inductive rule learning—see (Fürnkranz, 1999) for a survey—may be viewed as a general approach for combining local patterns into global classifiers. The basic idea is to repeatedly find the best local pattern and add it to a growing theory. The goodness of the local pattern is measured with some heuristic criterion that measures the deviation of the class distribution of the local pattern from the overall class distribution of the training examples. We will discuss such measures later on in this paper (Section 4). In the simplest case, local patterns are added until each example is covered by at least one local pattern.[1] Figure 1 shows the basic covering algorithm.

Rules are an obvious choice for local classifiers because a rule will typically only cover a subset of the entire example space. Consequently, rules are frequently used as a representation for local pattern discovery tasks such as association rule mining (Agrawal et al., 1995; Hipp et al., 2000) and subgroup discovery (Klösgen, 1996; Wrobel, 1997; Scheffer and Wrobel, 2002; Lavrač et al., 2004).

---

[1] In practice, this constraint is often relaxed to avoid overfitting.

However, the covering algorithm does not depend on rule-based local patterns. Ferri et al. (2004) elegantly generalized the covering framework to arbitrary classifiers by defining the locality via a confidence threshold: The classifier is trained on all training examples, but it will not issue a prediction unless it has a certain (user-specified) minimum confidence in its prediction. All training examples that are classified with this minimum confidence are then removed, and a new (possibly different type of) classifier is learned from the remaining examples.

A key problem for constructing a global theory out of local patterns is that the local patterns are discovered in isolation, whereas they will be used in the context of other patterns. The covering strategy partially addresses this problem by learning rules in order: all examples that are covered by previous patterns are removed from the training set before a new pattern is learned. This guarantees that the new local pattern will focus on new, unexplored territory. However, it also ignores the evidence contained in the removed examples, and the successive removal of training examples eventually leads to training sets with a very skewed class distribution, and possibly isolated, scattered examples.

As a remedy for this problem, several authors (Cohen and Singer, 1999; Weiss and Indurkhya, 2000; Gamberger and Lavrač, 2000) have independently proposed the use of *weighted covering* (Figure 2). The basic idea is to generalize the covering algorithm by introducing example weights. Initially, all examples have a weight of 1.0. However, the weights of examples that are covered by a rule will not be set to 0.0 (which is the equivalent to removing them from the training set), but instead their weight will only be reduced. This ensures that their influence on the evaluation of subsequent local patterns is reduced, but not entirely eliminated. Different algorithms use different weight adaptation formulas, ranging from error-based procedures motivated by boosting (Cohen and Singer, 1999) to simple techniques such as using the inverse of (one plus) the number of previous rules that cover the example (Gamberger and Lavrač, 2000).

Most weighted covering algorithms also adopt a very simple stopping criterion, namely they simply learn a fixed number of rules. Diversity of the rules is encouraged by the re-weighting of the examples, but it is no longer enforced that each example is covered by a rule. Also, the number of learned rules is typically higher, which has the effect that most examples will be covered by more than one rule. Thus, weighted covering algorithms have two complementary advantages: on the one hand they may learn better local pattern because the influence of previously covered patterns is reduced but they are not entirely ignored, on the other hand they will produce a better classifier by combining the evidence of more rules, thus exploiting the redundancy contained in an ensemble of diverse local patterns (Dietterich, 2000).

While covering and weighted covering try to take into account the context of previous patterns before generating a new local pattern, an alternative strategy is to try to "guess" what subsequent patterns may look like. One attempt into that direction is the PART algorithm (Frank and Witten, 1998), which does not

```
function WEIGHTEDCOVERING(Examples)

# initialize classifier and example weights
GlobalClassifier ← ∅
foreach Example ∈ Examples
    WEIGHT(Example) = 1.0

# loop for a fixed number of iterations
for i = 1...n

    # find the best local pattern
    LocalPattern ← FINDBESTLOCALPATTERN(Examples)

    # add the local pattern to the classifier
    GlobalClassifier ← GlobalClassifier ∪ LocalPattern

    # reduce the weight of covered examples
    REDUCEWEIGHTS(COVERED(LocalPattern,Examples))

return GlobalClassifier
```

Fig. 2. The Weighted Covering Algorithm

learn the next local pattern in isolation but (conceptually) learns a global model in the form of a decision tree. From this tree, a single path is selected as the next local pattern that can be added to the theory.[2] In essence, this idea is a special case of the delegating classifiers framework discussed above (Ferri et al., 2004).

The best local patterns are typically found via a heuristic search, using some heuristic evaluation metric as a guide. We will discuss a few such measures further below, but first we we have to explain coverage spaces.

3   Coverage Spaces

In recent work, Fürnkranz and Flach (2005) introduced the framework of *coverage spaces* for analyzing rule evaluation metrics. Coverage spaces are a quite similar to ROC-spaces, the main differences being that coverage spaces work with absolute numbers of true positives and false positives (covered positive and negative examples), whereas ROC-spaces work with true positive and false positive rates. A rule (or a rule set) that covers p out of a total of P positive examples and n out of N negative examples is represented as a point in coverage space with the co-ordinates (n, p).

Adding a rule to a rule set increases the coverage of the rule set because an additional rule can only add new examples to the set of examples that are covered by the rule set. All positive examples that are uniquely covered by the newly added rule contribute to an increase of the true positive rate on the training data.

---

[2] The implementation of this phase can be optimized so that the selected branch can be grown directly, without the need of growing an entire tree first.

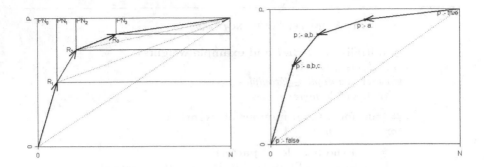

Fig. 3. Schematic depiction of the paths in coverage space for (left) the covering strategy of learning a rule set adding one rule at a time and (right) greedy specialization of a single rule.

Conversely, covering additional negative examples may be viewed as increasing the false positive rate on the training data. Therefore, adding rule r_{i+1} to rule set R_i effectively moves from point $R_i = (n_i, p_i)$ (corresponding to the number of negative and positive examples that are covered by previous rules), to a new point $R_{i+1} = (n_{i+1}, p_{i+1})$ (corresponding to the examples covered by the new rule set). Moreover, R_{i+1} will typically be closer to (N, P) and farther away from $(0, 0)$ than R_i.

Consequently, learning a rule set one rule at a time may be viewed as a path through coverage space, where each point on the path corresponds to the addition of a rule to the theory. Such a *coverage path* starts at $(0, 0)$, which corresponds to the empty theory that does not cover any examples. Figure 3 shows the coverage path for a theory with three rules. Each point R_i represents the rule set consisting of the first i rules. Adding a rule moves to a new point in coverage space, corresponding to a theory consisting of all rules that have been learned so far. Removing the covered examples has the effect of moving to a subspace of the original coverage space, using the last rule as the new origin. Thus the path may also be viewed as a sequence of nested coverage spaces PN_i. After the final rule has been learned, one can imagine adding yet another rule with a body that is always true. Adding such a rule has the effect that the theory now classifies *all* examples as positive, i.e., it will take us to the point $\tilde{R} = (N, P)$. Even this theory might be optimal under some cost assumptions.

For finding individual rules, the vast majority of algorithms use a heuristic top-down hill-climbing[3] or beam search strategy, i.e., they search the space of possible rules by successively specializing the current best rule

---

[3] If the term "top-down hill-climbing" sounds self-contradictory: hill-climbing refers to the process of greedily moving towards a (local) optimum of the evaluation function, whereas top-down refers to the fact that the search space is searched by successively specializing the candidate rules, thereby moving downwards in the generalization hierarchy induced by the rules.

(Fürnkranz, 1999). Rules are specialized by greedily adding the condition which promises the highest gain according to some *evaluation metric*. Just as with adding rules to a rule set, successive refinements of a rule describe a path trough coverage space (Figure 3, right). However, in this case, the path starts at the upper right corner (covering all positive and negative examples), and successively proceeds towards the origin (which would be a rule that is too specific to cover any example).

As we will see in the following, coverage spaces are well-suited for visualizing the behavior of evaluation metrics by looking at their *isometrics*, i.e., the lines that connect the rules that are evaluated equally by the used heuristic (Fürnkranz and Flach, 2005).

4   Rule Evaluation Measures

In each iteration, the covering algorithm needs to select the "best" local pattern that can be added. Informally, a good local pattern is a pattern for which the class distribution of the instances that it covers differs considerably from the overall class distribution. In a concept learning scenario (where we have only two classes, positive and negative examples for the target concept), we will try to identify regions of the instance space in which instances of the concept are denser than in the overall example distribution, i.e., in regions where there is a higher proportion of positive examples.

4.1   Trading Off Precision and Coverage

Numerous rule evaluation measures have been proposed in various contexts (Fürnkranz, 1999; Lavrač et al., 1999; Tan et al., 2002). In the following, we will concentrate on a family of well-known evaluation metrics for subgroup discovery (Klösgen, 1992). They have in common that they trade off two basic components:

Precision Gain $g = \frac{p}{p+n} - \frac{P}{P+N}$ is the difference between the proportion of positive examples in the examples covered by the local pattern and the overall proportion of positive examples.

Coverage $c = \frac{p+n}{P+N}$ is the proportion of all examples that are covered by the local pattern.

Figure 4 shows the isometrics in coverage space for these two basic heuristics. Note that the second term of precision gain is constant for all local patterns. Thus, maximizing precision gain is the same as maximizing precision, and the isometric structure of precision gain is the same as the one for precision itself: The rules with the lowest evaluation are those on the N-axis because they only cover negative examples. Here, precision has its minimum value of 0 and precision gain the minimum value of $-P/(P + N)$. The examples with the highest evaluation can be found on the P-axis because those are the ones that cover only positives examples. There, precision has its maximal value of 1, and precision gain the

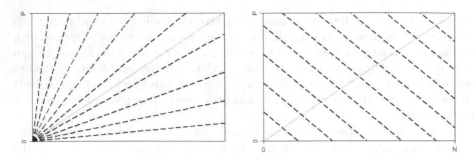

Fig. 4. Isometrics for precision gain (left) and coverage (right).

maximum value of $1 - P/(P + N)$. In between, the isometrics rotate around the point $(0,0)$, the empty rule. For example, all rules on the diagonal (those for which the covered positives and negative examples are distributed in the same way as the examples in the overall distribution) are evaluated in the same way with this heuristic (with the value 0 in the case of precision gain and with $P/(P+N)$ for precision). The isometrics of coverage move in parallel lines from the empty rule (no coverage) to the universal rule (covering all examples). The lines have an angle of 45 degrees with the N- and P-axes because for coverage there is no difference in importance for covering a positive or covering a negative example.

Klösgen (1996) identified three different variations for combining these two measures, which satisfy a set of four basic axioms proposed by Piatetsky-Shapiro (1991) and Major and Mangano (1995). He further showed that several other measures are equivalent to these. Wrobel (1997) added a fourth version. All four measures only differ in the way in which they trade off coverage c versus precision gain g. These measures are:

$$\text{(a) } \sqrt{c}g \qquad \text{(b) } cg \qquad \text{(c) } c^2g \qquad \text{(d) } \frac{c}{1-c}g$$

The isometrics of these measures are shown in Figure 5. Measure (a) was proposed by Klösgen (1992). Its idea is to perform a statistical test on the distribution of precision gain, under the assumption that, if the true precision of the rule were the same as the overall precision in the example set, the observed value for precision gain should follow a binomial distribution around 0. The variance of this distribution brings in the factor $\sqrt{c}$. The isometrics show that the measure has a slight tendency to prefer rules that are near the origin. In that region, the isometrics start to bend towards the origin, which means that rules with low coverage need smaller deviations from the diagonal than larger rules with the same evaluation.

Measure (b) is weighted relative accuracy, as proposed independently by Piatetsky-Shapiro (1991) and Lavrač et al. (1999). It has linear isometrics, parallel to the diagonal. Thus, all rules that have the same normal distance from the

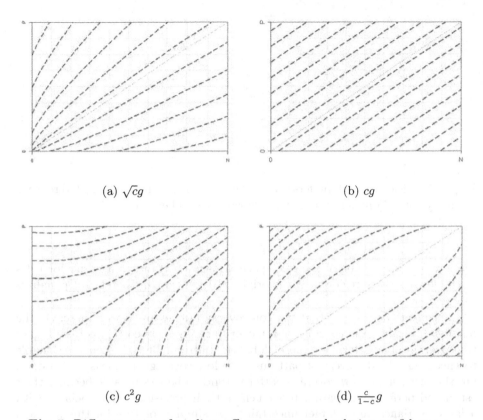

(a) $\sqrt{c}g$

(b) cg

(c) c^2g

(d) $\frac{c}{1-c}g$

Fig. 5. Different ways of trading off coverage c and relative confidence g.

diagonal are evaluated in the same way, independent of their location in coverage space. In comparison to (a), this has increased the influence of coverage, with the result that smaller rules are no longer preferred.

Wrobel (1997) proposed to further strengthen the influence of coverage by squaring it, resulting in measure (c). This results in an isometric landscape that has a clear tendency to avoid the region with low coverage near the lower left corner (see Figure 5, lower left). Obviously, the rules found with this measure will have a stronger bias towards generality.

Klösgen (1992) has shown that measure (d) is equivalent to several other measures that can be found in the literature, including a χ^2-test. It is quite similar to the first measure, but its edges are bent symmetrically, so that rules with high coverage are penalized in the same way as rules with a comparably low coverage.

It is quite interesting to see that in regions with higher coverage, the isometrics of all measures except (d) approach parallel lines, i.e., with increasing rule coverage, they converge towards some measure that is equivalent to weighted relative accuracy. However, measures (a), (b), and (c) differ in their behavior near

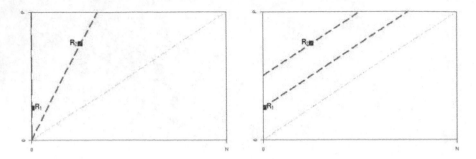

Fig. 6. Precision ($w = 0$) prefers the smaller, pure rule, whereas weighted relative accuracy ($w = 1$) prefers the larger rule with several exceptions

the low coverage region of the rule space. Measure (a) makes it easier for rules in the low-coverage region near the origin, (b) is neutral, whereas (c) penalizes this region.

It seems to be the case that two controversial forces are at work here: On the one hand, locality and coverage of a pattern are inversely correlated: the higher the coverage of a pattern, the more global the pattern. Thus, it seems reasonable to encourage the discovery of patterns with low coverage, as measure (a) does. On the other hand, low coverage patterns tend to be less reliable because their estimated performance parameters (such as their precision) are associated with a larger variance and a larger uncertainty. A simple, solution for this problem might be to try to avoid these regions if possible, as measure (c) does.[4] Weighted relative accuracy (b) tries to compromise between these two approaches. Note that precision may also be viewed in this framework, as giving no weight to the coverage of the rule (i.e., it is equivalent to $c^0 g$).

4.2   What Is the Optimal Trade-Off?

We have seen that all measures of the general form $c^w g$, for some $w \in \mathbb{R}, w \geq 0$, implement the basic idea of measuring the quality of a local pattern with the distance from the diagonal of the coverage space. The motivation for this approach is evident: the diagonal represents all rules that have the same overall distribution as can be found in the entire rule set, and the goal of local pattern discovery is to find a pattern that covers a set of instances that deviate significantly from this default distribution. The individual measures differ only in the way they measure this distance in different regions of coverage, i.e., in their different choices of w.

---

[4] This is related to the *small disjuncts problem*: rules with high coverage are responsible for a large part of the overall error of a rule set. Nevertheless, the experiments in (Holte et al., 1989) suggest that avoiding them entirely is not a good strategy.

Consider the example shown in Figure 6. It shows two rules: R_1 is a pure rule, covering one fourth of all positive examples and no negative examples. R_2, on the other hand, covers 3/4 of all positive examples but also a fourth of all negative examples, in a distribution where the prior probability of a positive example is 4/9. Thus, the precision gain of R_1 is $1 - 4/9 = 5/9 = 0.444$, whereas the precision gain of R_2 is $\frac{3/4 \times 4/9}{3/4 \times 4/9 + 1/4 \times 5/9} - 4/9 = 12/17 - 4/9 = 0.261$. Clearly, rule R_1 is better according to this criterion (cf. also the left graph of Figure 6).

On the other hand, if we evaluate with weighted relative accuracy, we get a different picture: rule R_1 covers one fourth of all positive examples, i.e. $1/4 \times 4/9 = 1/9$ of all examples. Rule R_2, on the other hand, covers a 3/4 of the positives, and 1/4 of the negatives, in total $3/4 \times 4/9 + 1/4 \times 5/9 = 17/36$ of the total number of examples. Thus, weighted relative accuracy, which multiplies coverage with precision gain, yields only $1/9 \times 1 = 1/9 = 9/81$ for rule R_1, but $17/36 \times (12/17 - 4/9) = 10/81$ for R_2. Note that these values are independent of the absolute number of examples that are covered by the rule, they only depend on the proportion of examples covered.[5]

However, intuitively, the validity and interestingness of the found patterns is not entirely clear. If rule R_1 covers only one or two positive examples, rule R_2 seems to be preferable because it is backed up with a larger amount of evidence and is therefore presumably more reliable. In our case, rule R_2 would cover about 3.5 examples, but it easy to construct examples where R_2 covers an arbitrary number of examples (increase the total number of examples and/or move the point upwards on its WRA isometric). On the other hand, if R_1 covers thousands of examples, a pure group of that size seems to be interesting irrespective of the total training set size.

Thus, we would propose that with growing coverage, coverage becomes less and less important for the evaluation of the quality of a found local pattern. In a crude form, this assumption can also be found in the support/confidence pruning framework that is paramount in association rule discovery: Rules below a given support threshold are not considered at all, rules above the given threshold are evaluated with their precision.[6] It is also related to trading off recall and precision. While the perfect value for such a trade-off is clearly application-dependent, we believe that for the case of discovering global patterns from local patterns, many pure rules are preferable to a few large but impure rules, provided it is established that the precision estimate of the rule is valid, i.e., it will generalize to uncovered examples. The latter, of course, is not true in typical rule learning applications, where the majority of the rules that are found with precision are rules covering only a few examples.

---

[5] This will not change if absolute coverage instead of relative coverage is used in the formula because a multiplication with a constant $(P + N)$ will not change the isometric structure for any given coverage space.

[6] Note, however, that support and coverage are not exactly the same: support is the proportion of covered *positive* examples, whereas coverage ist the proportion of *all* covered examples.

In any case, we believe that one of the main problems with the use of precision as a rule learning heuristic is that it is very susceptible to overfitting. Rules that cover only one or a few examples on the training set are evaluated with 100% precision, but their true precision in the entire domain will typically be much worse. Thus, many techniques have been proposed to make precision estimates more conservative, most prominently the Laplace- and m-estimates (Cestnik, 1990; Clark and Boswell, 1991). In the next section, we propose an alternative route that uses meta-learning for predicting the "true" precision of a rule.

5   Meta-learning Rule Precision

We have tried to motivate that precision (gain), which is of the form $c^w g$ for $w = 0$, may be a better evaluator for local patterns than other measures that use a value $w > 0$, provided that the precision values are not the result of overfitting. In this section, we show experiments for learning a function that predicts the rule's precision on an independent test set based on the rule's coverage on the training set. The basic idea is to generate a large number of rules, observe their precision on the training set and and independent test set, and learn a function that predicts the test set precision from the (absolute) number of covered examples on the training set. More details on this work can be found in (Fürnkranz, 2004a;b).

5.1   Meta Data Generation

We used a simple covering algorithm for learning a set of rules. For each learned rule, we recorded the numbers of covered positive and negative examples on both the training and an independent test set. We recorded these statistics not only for *final rules*—those rules that would be used in the final theory—but also for all their ancestors, i.e., for all *incomplete rules* that were eventually refined into a final rule. These can be simply obtained by deleting the final conditions of each rule. The main motivation for this step is that we want to have complete information on each path in the refinement graph that yields a final rule. Figure 7 shows the meta data generation algorithm in pseudo-code.

The algorithm for generating the individual rules is a straight-forward greedy top-down algorithm: rules are refined until no further refinement is possible. At each refinement step, all possible immediate refinements (adding one condition) are evaluated and the best one is selected. Among all rules encountered during this search, the best rule is eventually returned. Note, however, that the best rule need not be the last one searched.

We did not implement any method for pruning the obtained rules. Our main goal is to study the test set performance of individual rules, and not so much to learn a good theory. Therefore, the evaluation of possibly overfitting rules is very important to us. As a consequence, we chose not to implement any filtering heuristics which would prune those rules away. To ensure some variety in the size of the learned rules by using evaluation heuristics with very different biases (as

```
procedure GENERATERULES(TrainSet, TestSet)

# loop until all positive examples are covered
while POSITIVE(TrainSet) ≠ ∅

    # find the best rule
    Rule ← GREEDYTOPDOWN(TrainSet)

    # stop if it doesn't cover more pos than negs
    if |COVERED(Rule, POSITIVE(Examples))|
        ≤ |COVERED(Rule, NEGATIVE(Examples))|
    break

    # loop through all predecessors
    Pred ← Rule
    repeat

        # record the training and test coverage
        TrainP ← |COVERED(Rule,POSITIVE(TrainSet))|
        TrainN ← |COVERED(Rule,NEGATIVE(TrainSet))|
        TestP ← |COVERED(Rule,POSITIVE(TestSet))|
        TestN ← |COVERED(Rule,NEGATIVE(TestSet))|
        print Pred,Rule,TrainP,TrainN,TestP,TestN

        Pred ← REMOVELASTCONDITION(Pred)
    until Pred = null

    # remove covered training and test examples
    TrainSet ← TrainSet \ COVERED(Rule,TrainSet)
    TestSet ← TestSet \ COVERED(Rule,TestSet)
```

Fig. 7. Covering algorithm for generating and evaluating rules

will be explained below), some of which have a tendency to learn very general rules, while others are clearly prone to overfitting.

In order to collect statistics under a fairly broad set of conditions, we varied the following dimensions:

Datasets: We used 27 datasets with varying characteristics from the UCI repository. These datasets were selected because of their availability and moderate size. We did not include larger datasets (such as *shuttle*) because the region of interest (as we will see) is the region of rules with low coverage.

5x2 Cross-validation: For each dataset, we performed 5 iterations of a 2-fold cross-validation. 2-fold cross-validation was chosen because in this case the training and test sets have equal size, which makes a comparison of the obtained estimates easier. We collected statistics for all rules of all five iterations of two folds, i.e., a total of 10 per run.

Classes: For each dataset and each fold, we generated one dataset for each class, treating this class as the positive class and the union of all other classes as

Table 1. Search heuristics used in this study. p and n are the number of covered among a total of P and N positive and negative examples.

heuristic	formula
precision	$\frac{p}{p+n} \sim \frac{p-n}{p+n}$
Laplace	$\frac{p+1}{p+n+2}$
accuracy	$\frac{p+(N-n)}{P+N} \sim p - n$
weighted rel. acc.	$\frac{p+n}{P+N}\left(\frac{p}{p+n} - \frac{P}{P+N}\right) \sim \frac{p}{P} - \frac{n}{N}$
correlation	$\frac{p(N-n)-(P-p)n}{\sqrt{PN(p+n)(P-p+N-n)}}$

the negative class. Rules were learned for each of the resulting two-class datasets.

Heuristics: Finally, we ran the rule learner five times on each binary dataset, each time using a different search heuristic. We used the five heuristics shown in Table 1. The first four form a representative selection of search heuristics with linear isometrics (Fürnkranz and Flach, 2003), while the correlation heuristic (Fürnkranz, 1994) has non-linear isometrics. These heuristics represent a large variety of learning biases. For example, it is known that *WRA* and *Accuracy* tend to prefer simpler rules with high coverage, whereas *Precision* and *Laplace* tend to prefer possibly complex rules with high precision on the training set. Note that the correlation heuristic is equivalent to a χ^2-statistic (Fürnkranz and Flach, 2005), which in turn is equivalent to heuristic (d) of the previous section (Klösgen, 1992). We have not yet run experiments with heuristics (a) and (c).

In total, 5409 theories with 48,603 rules were learned. For all rules and ancestors we recorded their precision on the test set, resulting in statistics for a total of 114,375 rules. 13,399 rules did not cover any examples on the test set and were ignored. Our reasons for this were that on the one hand we did not have any training information for this rule (the test precision that we try to model is undefined for these rules), and that on the other hand such rules do not do any harm (they won't have an impact on test set accuracy as they do not classify any test example). Ignoring them seemed to be the most reasonable option for our purposes. The large majority of these ignored rules (9,806 rules) covered only a single positive and no negative examples on the training set. A total of 100,976 rules remained for the analysis.

Each rule is evaluated in the context of all previously learned rules, i.e., all examples covered by previous rules are removed from the dataset. Thus, later rules in a theory are learned from a smaller dataset than the first rules in the theory. This procedure was also mirrored in the test set. In other words, we assumed a decision list learning scenario, where an example is classified with the prediction of the first rule that fires on the example. Thus, rule n only receives examples (from both training or test sets) that are not covered by rules $1 \ldots n-1$.

5.2   Fitting Search Heuristics

We fitted several 2-dimensional functions to these meta data, with the goal of using them as a search heuristic inside the rule learner. We fitted the parameters of the following three types of heuristics:

- a neural network (fully connected with a five-node hidden layer, fitted using R's nnet procedure (Venables and Ripley, 2002))
- the m-estimate (Cestnik, 1990; Clark and Boswell, 1991), resulting in the function $\frac{p+1.6065*P/(P+N)}{p+n+1.6065}$ (fitted using R's nls procedure (Venables and Ripley, 2002))
- the generalized m-heuristic, which re-interprets the prior probability in the m-heuristic as a cost parameter (Fürnkranz and Flach, 2003), resulting in $\frac{p+0.785}{p+n+2.7153}$

The residual sum-of-squares showed the best fit for the m-heuristic ($rss = 7842.37$), followed by the neural network ($rss = 7897.1$) and the generalized m-heuristic ($rss = 8029.34$).

Table 2 shows the accuracy results (estimated by a 10-fold stratified cross-validation) for all eight heuristics[7] on the 27 data sets that were used for generating the meta data, as well as on 10 datasets that were not used in the training phase. At the bottom of Table 2, we also show the average rule sizes for each heuristic. As an independent benchmark, we also added the results of JRip, Weka's re-implementation of Ripper (Cohen, 1995), in two versions, without and with pruning. The results exhibit a fairly large variance. There are cases where the learned heuristics clearly outperform the five original heuristics (e.g., *labor*), but there are also cases where they are outperformed by at least one of them.

Table 3 shows the p-value of a paired t-test, and the number of wins and losses for each combination of a base heuristic with a meta-learned heuristic. It can be seen that the meta-learned heuristics outperform *Precision*, *Laplace*, and *Accuracy*. The differences for the neural network are not significant, but the trained m-estimates outperform these heuristics in all but one case at the 1% significance level. On the other hand, weighted relative accuracy and correlation, which correspond to heuristics (b) and (d) of Figure 5, are *en par* with the meta-learned heuristics. These two differ from the others in that they are symmetrical around the diagonal, i.e., they incorporate information about the prior probability of the problem. Among the meta-learned heuristics, only the m-heuristic takes this information into account.

In particular, the performance of the neural network is somewhat disappointing. Although it is the most expressive model class (the only one that could be trained to fit non-linear isometrics), the net did not surpass the results of its linear competitors, not even at a significance level of 5%. Overfitting could be one

---

[7] The neural network was implemented via a look-up table of the average prediction values of 10 different networks for all combinations of values $n \leq 50$ and $p \leq 50$. Precision was used for all larger values.

Table 2. Accuracy and number of learned rules for the five basic and three learned heuristics on 10 new datasets. For comparison, we also show the result of JRip without Pruning (-P) and JRip, and the average results of the algorithms on the 27 datasets that were used for training.

	Prec	Lap	Acc	WRA	Corr	NNet	MEst	GenM	JRip -P	JRip
anneal	99.00	99.00	98.75	98.62	98.75	99.37	99.37	99.25	98.50	97.62
audiology	76.55	74.78	80.97	85.84	80.53	77.43	77.43	76.11	73.89	72.12
breast-cancer	68.88	67.48	72.73	69.93	66.78	68.88	65.03	70.63	73.43	72.38
cleveland-heart	72.61	70.96	73.60	72.28	76.90	72.28	72.61	74.92	77.56	79.54
contact-lenses	66.67	62.50	66.67	83.33	66.67	70.83	70.83	70.83	70.83	75.00
credit	84.69	84.90	84.90	86.73	83.88	85.51	83.27	83.67	85.31	85.71
glass	59.81	57.94	62.62	59.35	65.89	58.88	61.22	59.35	66.36	69.16
glass2	74.23	73.62	78.53	76.07	81.60	73.62	77.91	76.69	81.60	79.14
hepatitis	79.35	81.29	78.71	78.71	78.06	76.13	77.42	80.00	81.29	79.35
horse-colic	74.18	71.20	79.35	84.24	82.07	75.82	75.82	75.00	78.80	84.78
hypothyroid	97.91	98.17	98.13	98.32	98.77	98.58	98.42	98.70	98.70	99.11
iris	95.33	96.00	92.00	92.00	92.67	92.00	96.00	95.33	90.67	95.33
krkp	99.06	99.28	97.25	94.34	98.22	99.09	99.25	99.28	99.44	99.09
labor	87.72	87.72	85.96	82.46	87.72	89.47	91.23	89.47	84.21	77.19
lymphography	82.43	82.43	77.70	79.73	79.05	81.08	81.08	81.76	74.32	81.08
monk1	78.23	81.45	79.84	81.45	81.45	73.39	81.45	81.45	84.68	89.52
monk2	47.93	49.11	47.93	54.44	55.03	53.25	48.52	51.48	49.11	52.07
monk3	82.79	86.07	78.69	77.05	74.59	87.70	83.61	85.25	81.97	86.07
mushroom	100.00	100.00	98.23	96.45	98.23	100.00	100.00	100.00	100.00	100.00
sick-euthyroid	95.86	96.17	96.74	96.30	96.68	95.98	96.27	96.24	96.68	97.72
soybean	88.73	89.17	91.07	87.26	90.19	90.78	91.22	90.63	91.07	90.19
tic-tac-toe	97.29	97.29	88.41	71.40	83.09	97.29	97.08	97.29	97.18	97.18
titanic	78.33	78.33	78.33	77.60	77.78	78.33	78.33	78.33	78.33	78.24
vote	94.48	94.94	94.48	94.94	94.02	95.86	95.17	94.71	95.40	96.32
vote-1	88.97	87.13	88.51	89.66	90.11	89.20	87.36	89.20	88.28	88.97
vowel	50.10	50.71	47.37	63.03	70.81	54.14	55.35	52.22	74.75	72.63
wine	92.13	92.13	93.82	95.51	93.82	93.82	92.70	93.26	94.38	91.57
average (27 old)	81.97	81.84	81.90	82.48	83.09	82.55	82.74	83.00	83.95	84.71
balance-scale	73.44	73.12	68.80	66.88	77.76	71.84	72.32	72.80	80.32	81.28
breast-w	94.85	94.85	95.28	94.28	95.57	95.14	94.56	95.14	93.71	95.14
credit-g	69.10	70.00	67.00	72.50	69.60	67.70	68.10	69.20	73.30	70.80
diabetes	68.23	69.66	69.27	71.88	69.01	70.31	71.88	68.75	72.92	74.22
ionosphere	93.45	94.30	89.46	89.74	88.03	94.02	93.73	94.02	90.60	88.60
primary-tumor	33.04	32.74	29.50	35.40	35.40	33.63	34.81	33.33	39.23	38.94
segment	91.39	90.61	88.10	92.29	94.94	91.64	91.77	91.17	95.76	95.11
sonar	62.02	63.46	68.75	67.79	73.08	66.83	65.87	67.31	77.40	76.44
vehicle	69.39	67.14	62.65	60.52	68.44	65.48	71.63	67.49	67.02	68.68
zoo	84.16	85.15	90.10	92.08	90.10	89.11	90.10	90.10	87.13	86.14
average (10 new)	73.91	74.10	72.89	74.34	76.19	74.57	75.48	74.93	77.74	77.53
avg. # rules (27 old)	40.41	36.93	32.56	4.74	13.63	30.22	30.81	30.85	14.11	8.11
avg. # rules (10 new)	83.20	78.30	86.50	4.30	21.20	64.20	67.60	68.40	18.50	9.80

cause, the above-mentioned absence of the prior probability as an additional input to the network another. Nevertheless, it is interesting to see how the network fitted the data. Figure 8 shows the surface of the learned evaluation function.

Table 3. Significance level of a paired t-test and number of wins and losses of pairwise comparisons between the base heuristics and the meta-learned heuristics.

	NNet	MEst	GenM
Precision	0.929 (10/23)	0.996 (9/25)	0.9996 (7/26)
Laplace	0.913 (10/23)	0.991 (12/21)	0.999 (10/22)
Accuracy	0.939 (13/21)	0.985 (13/22)	0.996 (12/23)
WRA	0.541 (20/15)	0.679 (16/19)	0.695 (16/19)
Correlation	0.178 (21/15)	0.291 (20/15)	0.321 (20/15)

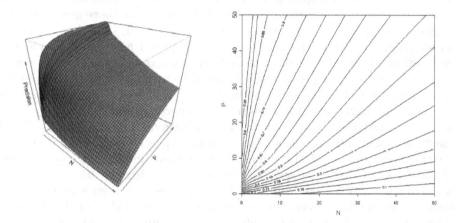

Fig. 8. Surface of a neural-net fit to the evaluation data

Fig. 9. Isometrics of a neural-net fit to the evaluation data

Note that the steep non-linear shape for low levels of N and P gradually shifts towards an almost linear shape. This is not surprising, as the bias of the training set precision can be expected to be much lower for rules with high coverage than for rules with low coverage, because it is easier to fit a small sample by chance. Figure 9 shows the isometrics of the learned neural network. It is quite obvious that the shape is very similar to the shape of precision, which would be lines rotating around the angle $(0,0)$. However, while the lines become increasingly straight the farther they move away from the origin, they are quite non-linear near the origin. In these regions, it might make a difference whether a rule is evaluated with precision on the training set or with the predicted test set precision. Moreover, the isometrics do not meet in the origin, but seem to rotate around some point below it. This is characteristic of the m-estimate, and related heuristics. and may partly explain the good performance of such heuristics.

In general, our results are on average somewhat below those of JRip, although there are numerous exceptions. This difference could have several reasons, among them differences in implementation (all other algorithms differed only in the used heuristics, whereas JRip is a completely independent implementation) and the fact that our algorithms did not use any kind of noise handling. A somewhat unexpected side result of our experiments is that the no pruning version of JRip often outperforms its pruning counter-part (17 wins vs. 18 losses, with a p-value of 0.86). Thus, it can be assumed that the lack of a pruning option does not necessarily hamper the performance of our simple implementation of the separate-and-conquer algorithm on this selection of datasets.

6   Conclusions

In our view, it is still an open question what functions should be used to evaluate candidate rules for local pattern discovery. In this paper, we have used the framework of coverage spaces to investigate a well-known family of evaluation metrics for subgroup discovery, which trades off coverage and precision of a rule. Our proposition is that for patterns with high coverage, coverage is of minor importance, and precision (or precision gain) should be used for evaluating the quality of the found pattern. The main motivation for including coverage seems to be the danger of overfitting. As an alternative to coverage-based overfitting avoidance, we investigated the possibility of meta-learning a function for predicting the "true" value of the precision of a rule. Our empirical results show that meta-learning improves over several commonly used evaluation metrics. However, two of the measures that were originally proposed for subgroup discovery were *en par* with the meta-learned heuristics. We take this as evidence that in future work, we should (a) concentrate on investigating the quality of this family of subgroup discovery measures for inductive rule learning, and (b) repeat some of our meta-learning experiments with this function family. In particular, we plan to use a meta-learning approach like the one reported in this paper for fitting the parameter w of the function family $c^w g$. A key difference is that these heuristics (like the m-estimate, which also performed quite well) take the prior class distribution into account. We expect that including this information as a third parameter will also improve the results of the neural network meta-heuristic. In effect, this means switching from meta-learning in conventional $2d$-ROC space to meta-learning in $3d$-ROC space (Flach, 2003).

Acknowledgments

I would like to thank the participants of the Dagstuhl workshop on Local Patterns for enlightening discussions on the subjects of this paper. I would also like to thank the many people that make great software freely available. Weka, R, Perl and Cygwin were invaluable for this work.

References

R. Agrawal, H. Mannila, R. Srikant, H. Toivonen, and A. I. Verkamo. Fast discovery of association rules. In U. M. Fayyad, G. Piatetsky-Shapiro, P. Smyth, and R. Uthurusamy, editors, *Advances in Knowledge Discovery and Data Mining*, pages 307–328. AAAI Press, 1995.

B. Cestnik. Estimating probabilities: A crucial task in Machine Learning. In L. Aiello, editor, *Proceedings of the 9th European Conference on Artificial Intelligence (ECAI-90)*, pages 147–150, Stockholm, Sweden, 1990. Pitman.

P. Clark and R. Boswell. Rule induction with CN2: Some recent improvements. In *Proceedings of the 5th European Working Session on Learning (EWSL-91)*, pages 151–163, Porto, Portugal, 1991. Springer-Verlag.

W. W. Cohen. Fast effective rule induction. In A. Prieditis and S. Russell, editors, *Proceedings of the 12th International Conference on Machine Learning (ML-95)*, pages 115–123, Lake Tahoe, CA, 1995. Morgan Kaufmann.

W. W. Cohen and Y. Singer. A simple, fast, and effective rule learner. In *Proceedings of the 16th National Conference on Artificial Intelligence (AAAI-99)*, pages 335–342, Menlo Park, CA, 1999. AAAI/MIT Press.

T. G. Dietterich. Ensemble methods in machine learning. In J. Kittler and F. Roli, editors, *First International Workshop on Multiple Classifier Systems*, pages 1–15. Springer-Verlag, 2000.

C. Ferri, P. Flach, and J. Hernández. Delegating classifiers. In R. Greiner and D. Schuurmans, editors, *Proceedings of the 21st International Conference on Machine Learning (ICML-04)*, pages 289–296, Sydney, Australia, 2004. Omnipress.

P. A. Flach. The geometry of ROC space: Using ROC isometrics to understand machine learning metrics. In T. Fawcett and N. Mishra, editors, *Proceedings of the 20th International Conference on Machine Learning (ICML-03)*, pages 194–201, Washington, DC, 2003. AAAI Press.

E. Frank and I. H. Witten. Generating accurate rule sets without global optimization. In J. Shavlik, editor, *Proceedings of the 15th International Conference on Machine Learning (ICML-98)*, pages 144–151, Madison, Wisconsin, 1998. Morgan Kaufmann.

J. Fürnkranz. FOSSIL: A robust relational learner. In F. Bergadano and L. De Raedt, editors, *Proceedings of the 7th European Conference on Machine Learning (ECML-94)*, pages 122–137, Catania, Italy, 1994. Springer-Verlag.

J. Fürnkranz. Separate-and-conquer rule learning. *Artificial Intelligence Review*, 13 (1):3–54, February 1999.

J. Fürnkranz. Modeling rule precision. In J. Fürnkranz, editor, *Proceedings of the ECML/PKDD-04 Workshop on Advances in Inductive Rule Learning*, pages 30–45, Pisa, Italy, 2004a.

J. Fürnkranz. Modeling rule precision. In A. Abecker, S. Bickel, U. Brefeld, I. Drost, N. Henze, O. Herden, M. Minor, T. Scheffer, L. Stojanovic, and S. Weibelzahl, editors, *Lernen – Wissensentdeckung — Adaptivität. Proceedings of the LWA-04 Workshops*, pages 147–154, Humboldt-Universität zu Berlin, 2004b.

J. Fürnkranz and P. Flach. An analysis of rule evaluation metrics. In T. Fawcett and N. Mishra, editors, *Proceedings of the 20th International Conference on Machine Learning (ICML-03)*, pages 202–209, Washington, DC, 2003. AAAI Press.

J. Fürnkranz and P. Flach. ROC 'n' rule learning – Towards a better understanding of covering algorithms. *Machine Learning* 58(1):39–77, 2005.

D. Gamberger and N. Lavrač. Confirmation rule sets. In D. A. Zighed, J. Komorowski, and J. Żytkow, editors, *Proceedings of the 4th European Conference on Principles of Data Mining and Knowledge Discovery (PKDD-00)*, volume 1910 of *Lecture Notes in Artificial Intelligence*, pages 34–43, Lyon, France, September 2000. Springer, Berlin.

J. Hipp, U. Güntzer, and G. Nakhaeizadeh. Algorithms for association rule mining – a general survey and comparison. *SIGKDD explorations*, 2(1):58–64, June 2000.

R. Holte, L. Acker, and B. Porter. Concept learning and the problem of small disjuncts. In *Proceedings of the 11th International Joint Conference on Artificial Intelligence (IJCAI-89)*, pages 813–818, Detroit, MI, 1989. Morgan Kaufmann.

W. Klösgen. Problems for knowledge discovery in databases and their treatment in the statistics interpreter EXPLORA. *International Journal of Intelligent Systems*, 7(7):649–673, 1992.

W. Klösgen. Explora: A multipattern and multistrategy discovery assistant. In U. M. Fayyad, G. Piatetsky-Shapiro, P. Smyth, and R. Uthurusamy, editors, *Advances in Knowledge Discovery and Data Mining*, chapter 10, pages 249–271. AAAI Press, 1996.

N. Lavrač, P. Flach, and B. Zupan. Rule evaluation measures: A unifying view. In S. Džeroski and P. Flach, editors, *Proceedings of the 9th International Workshop on Inductive Logic Programming (ILP-99)*, pages 174–185. Springer-Verlag, 1999.

N. Lavrač, B. Kavšek, P. Flach, and L. Todorovski. Subgroup discovery with CN2-SD. *Journal of Machine Learning Research*, 5:153–188, 2004.

J. A. Major and J. J. Mangano. Selecting among rules induced from a hurricane database. *Journal of Intelligent Information Systems*, 4(1):39–52, 1995.

G. Piatetsky-Shapiro. Discovery, analysis, and presentation of strong rules. In G. Piatetsky-Shapiro and W. J. Frawley, editors, *Knowledge Discovery in Databases*, pages 229–248. MIT Press, 1991.

T. Scheffer and S. Wrobel. Finding the most interesting patterns in a database quickly by using sequential sampling. *Journal of Machine Learning Research*, 3:833–862, 2002.

P.-N. Tan, V. Kumar, and J. Srivastava. Selecting the right interestingness measure for association patterns. In *Proceedings of the 8th ACM SIGKDD International Conference on Knowledge Discovery and Data Mining (KDD-02)*, pages 32–41, Edmonton, Alberta, 2002.

W. N. Venables and B. D. Ripley. *Modern Applied Statistics with S*. Springer, fourth edition, 2002.

S. M. Weiss and N. Indurkhya. Lightweight rule induction. In P. Langley, editor, *Proceedings of the 17th International Conference on Machine Learning (ICML-2000)*, pages 1135–1142, Stanford, CA, 2000.

S. Wrobel. An algorithm for multi-relational discovery of subgroups. In *Proceedings of the First European Symposion on Principles of Data Mining and Knowledge Discovery (PKDD-97)*, pages 78–87, Berlin, 1997. Springer-Verlag.

Pattern Discovery Tools for Detecting Cheating in Student Coursework

David J. Hand, Niall M. Adams, and Nick A. Heard

Department of Mathematics
Imperial College London
{d.j.hand, n.adams, n.heard}@imperial.ac.uk

Abstract. Students sometimes cheat. In particular, they sometimes copy coursework assignments from each other. Such copying is occasionally detected by the markers, since the copied script and the original will be unusually similar. However, one cannot rely on such subjective assessment – perhaps there are many scripts or perhaps the student has sought to disguise the copying by changing words or other aspects of the answers. We describe an attempt to develop a pattern discovery method for detecting cheating, based on measures of the similarities between scripts, where similarity is defined in syntactic rather than semantic terms. This problem differs from many other pattern discovery problems because the peaks will typically be very low: normally only one or two cheating students will copy from any given other student.

1 Introduction

The word 'pattern' is used in this paper to describe an unexpected local peak in a probability function. Evidence for such an anomaly will come from a corresponding anomaly in a data set – that is, a local, unusually dense, accumulation of data points relative to some background model (1). This type of pattern arises in various contexts. The context with which this paper is concerned is that of assessing students' coursework, and we are concerned with detecting copying, which manifests itself in some scripts being unexpectedly similar. That is, the data points representing the copies are closer than expected, in some appropriate data space.

Pattern discovery involves two complementary aspects. On the one hand, it is necessary to identify potential patterns: to find the unusually high local density of data points. And, on the other, it is necessary to decide whether that 'unusually high local density' is such as to be statistically significant: with large data sets one must expect spurious congregations of data points to happen by chance, and one would like some measure of this probability.

The literature on pattern discovery has tended to focus on the first of these aspects. This is perhaps not surprising. For various reasons (discussed, for example, in (2)), most of the work in pattern discovery has occurred in computational disciplines. And computational disciplines are concerned with algorithms. Once a potential pattern has been found one can, at least in principle, hand things over

K. Morik et al. (Eds.): Local Pattern Detection, LNAI 3539, pp. 39–52, 2005.

to a domain expert to decide whether that potential pattern is real, and, if so, whether it is of any consequence. That this is often only possible in principle, and not in practice, is illustrated by the fact that such pattern discovery algorithms can easily throw up many thousands of potential patterns. It is unrealistic to expect any expert to sift through these, giving proper critical thought to each. The alternative, of setting the threshold for detection so that only the most striking data anomalies are flagged, is likely to indicate only those anomalies which are already well known. The problems with which this paper is concerned involve relatively small data sets, so that algorithmic efficiency is not an important issue.

The literature on the second aspect is much sparser and again there are sound reasons for this. This second aspect is fundamentally inferential, so that one might expect it to have been explored and developed within the primary discipline concerned with inference, namely statistics. However, as a discipline statistics has its roots in the first half of the twentieth century, prior to the development of computers. This restricted the size of data sets which could be handled, and pattern discovery is much less feasible in small data sets. For example, with a million data points, a small pattern (0.01% of the data, say) involves 100 points – a perfectly respectable number. However, with 1000 data points, 0.01% of the data involves less than one point. The chance of detecting such a structure as an anomaly is very small. Despite this, in recent years some formal inferential methods have been developed, and these are discussed in Section 3. As we have already noted, the particular type of pattern discovery addressed in this paper is rather unusual – it involves data sets which typically have relatively few cases, though they may have large dimensionality. This last factor may have deterred detailed statistical investigation in the past.

Cheating by a minority of dishonest students is a perennial issue. It is perhaps more prevalent in coursework rather than examinations or tests, simply because it is easier to perpetrate and more difficult to detect. In recent years, with the advent of the world wide web, the particular type of cheating called plagiarism, in which students pass off others' work as their own, has become especially important, because it is now so easy for dishonest students to locate material on the web and copy it. Because of this, search algorithms have been developed which trawl the web and compare a student's work with the results of this trawl. Plagiarism is a particular problem with open-ended questions – 'write an essay on X', or 'describe the major influences on Y', for example. In contrast, in closed questions, copying is perhaps a more critical form of cheating. In this paper we are concerned with mathematical questions, which might be of the form 'calculate the probability of some event', 'find the form of a function', or 'prove that some relationship holds', for example. Here students might be tempted simply to copy a solution from other students.

If there are not too many scripts, someone marking them may detect what seems to be a striking degree of similarity between scripts, and suspect collusion or copying. However, such suspicions, while serving as a basis for close monitoring of the students in the future, can hardly form a solid foundation for action. Something more formal is needed, something which can actually give a probability to the similarity, in the manner of a DNA test in crime detection. This pattern discovery problem has some rather unusual features. First, of course, it is not simply a pattern *matching* problem, as is DNA matching, in which one has the specimen found at the scene of the crime and the aim is to find a match in a database. It is a pattern *discovery* problem, which involves calculating and making inferences on *all* small clusters of objects. This leads to a combinatorially more complex problem (instead of matching 1 with n, yielding n possible matches, we must examine ^nC_r possible matches if we are seeking possible clusters of size r from a dataset of size n). Secondly, unlike most other pattern discovery problems, our 'local peaks' are likely to be represented by only a very few data points. If we had a class of 100, it may be quite likely that two friends have colluded, but it is unlikely that 20 have colluded. In a more standard pattern discovery problem, we will also be interested in the case of 20 similar points – so that we have to consider arbitrary r in the ^nC_r above. In fact, for our problem, we can restrict ourselves to searching for *pairs* of scripts which are strikingly similar.

In what follows, Section 2 describes the background and the data we have been working with. Section 3 discusses the inferential issues which are central to this problem, and describes our solutions. Section 4 presents our results. Finally, Section 5 draws some conclusions.

2 The Data

Our work in this area was motivated by suspiciously similar scripts in mathematics and statistics coursework. Appendix 6 shows a question and solution similar to those which motivated this work, and each script would have involved several such questions – say ten. All students will have answered, or attempted to answer, the same questions.

In order to apply formal inferential methods, it is necessary to represent the answers in a convenient representation. The most basic representation is the total mark a student obtains for the work. However, it is obvious that this does not yield a fine enough gradation for the data space to permit effective detection of cheating: with a total of 100 marks, and 100 students, we should expect many pairs of students to have very similar or even identical scores. Moreover, the chance of at least one pair having very close scores is very high (indeed, with 100 marks and 102 students, this probability is 1). It is necessary to expand the data space, by finding more detailed descriptions of the mark patterns. For example, with ten questions, each with ten marks, one might use the 10-component mark vector (this has 11^{10} possible configurations, instead of merely 101). This idea can be taken further, down to the level of the number of marks awarded to each part of each question. However, the parts of each question are likely to have only a few marks each (after all, if each question has only 10 marks, there is not much flexibility). Overall, this limits the fineness of the gradation which can be achieved in practice, and this, in turn (see Section 3) limits the precision of the inferences which can be made.

Most of the work on detecting plagiarism, and, indeed, other bibliometric work, has been based on matching occurrences of words and word counts. However, some work has been based on structural markers other than semantic terms. For example, in text, the structural markers might be punctuation marks. A recent example illustrates how effective this can be:

'In February 2003 a Cambridge politics lecturer named Glen Rangwala received a copy of the British government's most recent dossier on Iraq. He quickly recognised in it the wholesale copying of a twelve-year-old thesis by American doctoral student Ibrahim al-Marashi, "reproduced word for word, misplaced comma for misplaced comma". ... Rangwala noticed there were some changes to the original, such as the word "terrorists" substituted for "opposition groups", but otherwise much of it was identical. In publishing his findings, he wrote:
Even the typographical errors and anomalous uses of grammar are incorporated into the Downing Street document. For example, Marashi had written:
"Saddam appointed, Sabir 'Abd al-'Aziz al-Duri as head"...
Note the misplaced comma. The UK officials who used Marashi's text hadn't. Thus, on page 13, the British dossier incorporates the same misplaced comma:
"Saddam appointed, Sabir 'Abd al-'Aziz al-Duri as head"...'

Truss (3, p. 202)

We explored a similar structural description of the scripts. Many of the terms in a mathematical argument are arbitrary: one might use x instead of y, or α instead of β, for example. But other terms are universal: $+$, $-$, and $=$, in particular, cannot be written in other forms. Note that the other arithmetic operations of $\times$ and $\div$ do have alternative forms. Moreover, in general, the solutions to our questions had the scope for multiple usages of $+$, $-$, and $=$. This meant that a much finer gradation was possible if we coded the solutions in terms of these syntactic attributes. An extract from our data matrix is given in Appendix 7.

We suspect that students who attempt to disguise the fact that they have copied are unlikely to go to the extent of substantially altering the occurrences of the syntactic $+$, $-$, and $=$ terms, so that we believe the counts of these will serve as good descriptors for our purposes. There were 50 students in the class and, using these syntactic counts as descriptors, the data space was spanned by 60 component feature vectors. A priori, this might be expected to give substantial scope for inferential methods: in such a high-dimensional space identical responses might be unlikely to occur by chance.

3 Chance Similarity?

3.1 Scan Statistics

The ideas and tools of *scan statistics* (4) provide a basis on which we can decide whether an apparent pattern, suggested by an unexpected confluence of data points, represents a real structure in the underlying distribution. Unfortunately, this is a relatively new area, and most of the work has focussed on the one or two dimensional cases. This subsection summarises the ideas of scan statistics and the next explores approaches which may be suitable for our particular problem.

Suppose that we observe values of a single variable. A natural question is whether there is a tendency for certain values to occur more often than might be expected according to some background model. For example, if it is thought that the values should be uniformly distributed, is there a tendency for values to group, or if the values are generated by a point process, can it be modelled by a Poisson process, or is there evidence of clumping? Real examples include the possibility of illness clustering in time, of faults clustering because of a common cause, or of police deaths in the line of duty. An appropriate statistic to detect such clustering can be defined in terms of a small window, moving over the data space, with a count being made of the number of events it covers at each position. Of particular interest is the distribution of the maximum number of points covered by a window of given width, or the distribution of the width of the window required to cover a given number of points.

As a simple example, consider a sequence of N independent binary random variables, X_i, $i = 1, ..., N$. Let our null hypothesis be that these all have the same probability of taking value 1, so that $H_0 : X_i \sim Bern(p_0)$, $i = 1, 2..., N$, where $Bern(p)$ is the Bernoulli distribution with parameter p. Let our alternative hypothesis be $H_1 : X_i \sim Bern(p_1)$ for some $i = s, s + 1..., s + w - 1$, and $X_i \sim Bern(p_0)$ otherwise. Now define

$$Y_t = \sum_{i=t}^{t+w-1} X_i.$$

Then the *scan statistic* is

$$S_w = \max_{1 \leq t \leq N-w+1} Y_t.$$

Analogously, we will be concerned with the distribution of W_k, the shortest time period containing k events. Note that

$$P(W_k > w) = P(S_w < k).$$

In general, the distribution of W_k or S_w is difficult to find because of the dependencies involved: each window overlaps several of those before and after. The difficulty is illustrated by the following simple example, involving N observations from a uniform distribution on the unit interval:

$$X_i, \ i = 1, ..., N \ \ iid \ \sim f(x) = \begin{cases} x & 0 \le x \le 1 \\ 0 & else \end{cases}$$

Denote the order statistics by $X_{(i)}$ and let $W_k = \min\limits_{1 \le i \le N-k+1} \left[X_{(i+k-1)} - X_{(i)} \right]$.

Now,

$$P(W_k > w) = P\left(\min\limits_{1 \le i \le N-k+1} \left[X_{(i+k-1)} - X_{(i)} \right] > w \right)$$
$$= P\left(\bigcap_{i=1}^{N-k+1} \left(X_{(i+k-1)} - X_{(i)} > w \right) \right)$$

so that, to determine the distribution of W_k, all we have to do is to integrate the distribution for order statistics from a uniform distribution, $N!$ for $X_{(1)} \le X_{(2)} \le ... \le X_{(N)}$, over the domain $\left(X_{(i+k-1)} - X_{(i)} > w \right)$ for $i = 1, 2, ..., N - k+1$. Unfortunately, although the argument being integrated is straightforward enough, the region of integration is typically very complicated, making explicit integration impracticable except in certain special cases (one of which we discuss below).

Because of this difficulty, various approximations have been developed, two of which we describe below.

Product approximations are based on decomposing

$$P(W_k > w) = P\left(\bigcap_{i=1}^{N-k+1} \left(X_{(i+k-1)} - X_{(i)} > w \right) \right)$$

in terms of a Markov chain and approximating each term of this chain. For example, suppose we have a point process on the interval $[0, T)$, and define a moving window $[t, t+w)$, $0 \le t \le T - w$. For simplicity, let $T = Lw$, L an integer, and define Q_V to be the probability that there is no window of length w in V which contains as many as k points.

Let E_i denote the event that the interval $[(i-1)w, (i+1)w]$ does not include a window of length w containing as many as k points.

Then

$$Q_T = P(E_1) P(E_2|E_1) P(E_3|E_2 \cap E_1) ... P\left(E_{L-1} \Big| \bigcap_{i=1}^{L-2} E_i \right)$$

If we now assume that there are no long range relationship, then

$$P(E_i|E_{i-1} \cap ...) = P(E_i|E_{i-1})$$

If we also assume stationarity, then, further,

$$P(E_i|E_{i-1}) = P(E_j|E_{j-1}) = P(E_j \cap E_{j-1})/P(E_{j-1}) = Q_{3w}/Q_{2w},$$

from which

$$Q_T \approx Q_{2w} \left(\frac{Q_{3w}}{Q_{2w}} \right)^{T/w-2}$$

Better approximations can be made by increasing the range of the dependence, using for example, $P(E_i|E_{i-1} \cap E_{i-2} \cap \ldots) = P(E_i|E_{i-1} \cap E_{i-2})$.

As a second example, consider a Poisson process $\{X_t, t \geq 0\}$ with intensity λ. Define the *scanning process* $\{Y_t(w), t \geq 0\}$ with $Y_t = X_{t+w} - X_t$. Then the scan statistic with window w is $S_w = S_w(\lambda, T) = \max_{0 \leq t \leq T-w} Y_t(w)$. Without loss of generality, we can work with $T' = T/w$ and $\lambda' = \lambda w$. Now Y is an integer-valued stationary process with jumps ± 1. Let M_k be the number of times Y hits k in interval $(0, T-w)$. Then

$$\begin{aligned} P(W_k > w) &= P(S_w < k) \\ &= P(Y_0 < k \cap M_k = 0) \\ &\approx P(Y_0 < k) P(M_k = 0) \\ &= F_p(k-1; \lambda') P(M_k = 0) \end{aligned}$$

where $F_p(k; \mu)$ is the cdf of a Poisson distribution with mean μ

Using properties of Poisson processes, and where $p(k; \mu)$ is the pmf of this Poisson distribution, we obtain

$$P(W_k > w) \approx F_p(k-1; \lambda') \exp \left\{ - \left(1 - \frac{\lambda'}{k} \right) \lambda' (T' - 1) p(k-1; \lambda') \right\}$$

3.2 Applications in Cheating

As described above, it will be sufficient for us to detect pairs of scripts which are improbably similar. Our initial attempts were based on summarising the scripts by means of single values. These could be overall marks, or (and, as we will see, this is better) could be sums of the values derived by the syntactic description above. For convenience in this section, we shall call this overall score the student's score. We wish to know how probable it is that we should obtain a difference between the two closest scores as close or closer than that observed. If this probability is very small, we are justified in being suspicious about the two close scores. That is, we want to find

$$P(W_2 \leq w) = 1 - P(W_2 > w)$$

With N students, $P(W_2 > w)$ is given by the integral of $N!$ over the region defined by

$$0 \leq x_{(1)} \leq x_{(2)} \leq \ldots \leq x_{(N)} \leq 1 \text{ and } x_{(i+1)} - x_{(i)} > w \text{ for } i = 1, \ldots, N-1. \text{ It}$$
is not difficult to show that

$$P(W_2 > w) = \int_{(N-1)w}^{1} \int_{(N-2)w}^{x_N - 1} \ldots \int_{2w}^{x_4 - w} \int_{w}^{x_3 - w} \int_{0}^{x_2 - w} N! \, dx_1 dx_2 dx_3 \ldots dx_N$$

which, by the change of variables

$$
\begin{aligned}
y_1 &= x_1 \\
y_2 &= x_2 - w \\
y_3 &= x_3 - w \\
&\cdots\cdots\cdots\cdots \\
y_N &= x_N - (N-1)\,w
\end{aligned}
$$

can be written as

$$
\int_0^{1-(N-1)w} \int_0^{y_N} \int_0^{y_{N-1}} \cdots \int_0^{y_2} N!\,dy_1 dy_2 dy_3 ... dy_N
$$

which evaluates to

$$
P\left(W_2 > w\right) = \begin{cases} \{1 - (N-1)\,w\}^N & for \ w \le (N-1)^{-1} \\ 0 & for \ w > (N-1)^{-1} \end{cases} \tag{1}
$$

The $w > (N-1)^{-1}$ case is uninteresting, of course.

The difficulty of using this in practice is illustrated by the following examples. Suppose that the overall score can take values in the range 0 to r. Then, under the null hypothesis of independent uniform sampling from $0, ..., r$ the probability that all N scores differ is $(r+1)! \big/ (r+1)^N (r-N+1)!$ which is relatively small unless r is very large relative to N. For example, with $N = 20$ and $r = 99$ this probability is 0.13. In order to reduce it to a value which would give one confidence that something untoward had occurred one would need to increase the range of possible scores substantially. Moreover, this calculation has been based on assuming a uniform distribution over the range of the scores. Since, in fact, the scores are likely to follow a non-uniform distribution, the situation is even worse than these calculations suggest.

Using these figures of $N = 20$ and $r = 99$ in expression (1) above and using the uniform approximation to the distribution of scores over the range $0, ..., 99$ gives $P\left(W_2 \le 0.02\right) = 0.999993$. That is, we are almost certain to observe two scores within 2 points of each other if we have 20 students, each scoring in the range $0, ..., 99$. Figure 1 shows a plot of the value of r, the range of the possible scores needed, in order that a difference of 2 points would arise with probability 0.01, for given values of N, the number of students in the class, making the conservative assumption of a uniform score distribution.

Approaches such as this, in which we calculate an 'overall score' for the students, are sacrificing information: they are reducing the vector of values to a single summary statistic. This suggests that we can do better. In particular, for example, we could examine the distances between the vectors of student scores, not in terms of a single unidimensional summary (the overall score) but using a formal distance measure. We could then calculate the probability of observing a minimum distance smaller than that observed. Unfortunately, however, we cannot use simple order statistics on the distances for this, because such distances

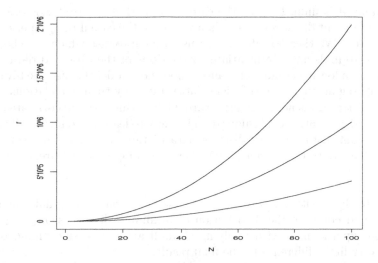

Fig. 1. Range of marks, r, needed to ensure that a difference of 2 marks will occur with a probability of less than 0.01 (top curve), 0.02, and 0.05 (bottom curve) when there are N students.

are not independent. Moreover, derivations will be difficult because the underlying distribution which we should use as the null hypothesis – the distribution we assume for the distances if there is no grouping arising from cheating – is generally unknown and certainly not uniform. For these reasons, Monte Carlo methods seem to be the best approach. These are based on fitting a parametric model to the joint observed distribution of student's vector of marks, and then generating new data sets, of the same size, from this model. The relevant probability is the proportion of these data sets which have a smallest pairwise distance as small or smaller than that actually observed. The model does not have to be too accurate, since slight inaccuracies will not have a dramatic effect on the distribution of inter-point distances. Note, however, that one should try to avoid fitting the distribution of each component of the mark vector as independent components: the likely correlation between the components will mean that a model which assumes independence will underestimate the probability of achieving small pairwise distances. That is, an independence model is likely to lead to exaggerating the apparent extremeness of observed small distances.

Our aim, then, is to generate samples of size 50 from the same 60-dimensional multivariate discrete distribution as the observed data, and record the proportion of these samples which have a minimum interpoint distance less than that observed in the actual data. To do this, we need a model for the 'distribution of the observed data'. In principle, such a model could be found by fitting a log-linear model to the data, but in practice, with only 50 data points and with 60 variables this would be impracticable. As we have already noted, it would also be unwise to go to the other extreme, and model the variables as independent, partly because it is unlikely to be the case, and partly because it would lead to minimum interpoint distances larger than would be the case if the proper dependence was used. We therefore decided to use a compromise which matched the univariate marginals and the bivariate correlations of the observed data. Note that this is not a log-linear model because it does not model the discrete bivariate distributions (again there is insufficient data), but only their correlations.

Even this strategy is not straightforward. Generating discrete correlated data is difficult, and generating continuous data is likely to lead to too few small values of the minimum distance (with discrete data distances of zero can occur, but with continuous data they cannot). We therefore adopted an indirect strategy, as follows:

1. Generate data with the observed discrete univariate marginals independently, and compute the distribution of minimum distances, m_1. (In fact, we also experimented with applying a small amount of smoothing, but it made very little difference to the final result.)
2. Generate data from independent normal distributions, with means and variances matching those of the observed variables. Discretise these data to the ranges of the observed variables. Calculate the mean minimum distance, $\bar{m}_2$.
3. As (2), but with correlated multivariate normal data (correlated multivariate normal data are easy to generate), matching the observed correlations. Again discretise the data and again calculate the mean minimum distance, $\bar{m}_3$.
4. Adjust m_1 to take account of the correlations, yielding $m_1' = m_1 \times \bar{m}_2/\bar{m}_3$.

This strategy is clearly fairly crude, and there are obvious improvements which could be made. However, it is worth bearing in mind that the overall model does not have to be too accurate – the distribution of minimum distances is unlikely to be substantially affected by small differences.

4   Results

Ten of the 60 columns of data showed constant zeros, so we dropped these from our analysis, leaving 50 records, each measured on 50 variables. The correction ratio $\bar{m}_2/\bar{m}_3$ was 0.845. Applying this value to the observed minimum distances in 5000 simulations of 50 data points led to the distribution shown in Figure 2. The arrow to the left of the diagram shows the minimum distance in the real data. The probability of observing such an extreme value from this distribution is vanishingly small. We are right to be highly suspicious of the two scripts which led to this distance.

5   Conclusion

With the advent of the internet, plagiarism, in which students download essays and coursework materials from the web, passing it off as their own work, has

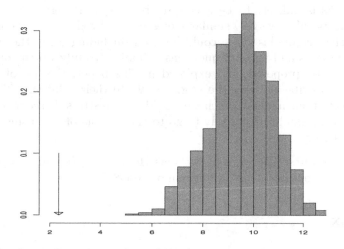

Fig. 2. Simulated minimum distance distribution, and observed (arrow) minimum distance in the real data.

generated increasing concern. But simple copying of coursework from other students is also an issue. Traditionally, detecting copying has relied on the markers, but this is unreliable and subjective. It is unreliable because, with $n(n-1)/2$ pairs of scripts amongst n students, it is all too easy to miss a very similar pair: unless they are marked consecutively, one is unlikely to notice similarities, perhaps especially because copied scripts may well be correct scripts (it seems rather pointless copying a script one cannot be sure is fairly good). It is subjective because the extent of 'similarity' will lie in the mind of the marker. What is needed is some more formal approach, and in particular an approach which allows one to attach a measure of probability to the similarity between scripts. In this paper we have described an attempt to construct just such an approach and measure. We have constructed this approach in the context of a broader theory of *pattern discovery* which we have been developing (Hand and Bolton, 2004).

The analysis presented in this paper is far from perfect. Perhaps its major shortcoming arises from the difficulty of generating the distribution of minimum distances. Our approach is crude, and could doubtless be improved using the tools of modern Monte Carlo methods – although it is not straightforward because of the discrete nature and high dimensionality (relative to a small sample size) of the data. In any case, we believe that the model from which the simulated minimum distances are generated need not be too accurate a model of the underlying distribution (under the null hypothesis of no copying). We believe that copying is likely to result in a very small similarity measure – far out in the tail of the distribution. This was certainly the case in the example we gave.

Two other aspects of the method we have described are worth stressing. The first is the need to produce a very large potential data space. Simple use of total marks is unlikely to be successful because it is very likely that two honest students will have very similar scores purely by chance. Various ways of increasing the size of this space could be used, including using the individual marks of questions and parts of questions. This is certainly worth considering. However, we also proposed, and explored in this paper, the use of syntactic measures to describe the way the students wrote their solutions. This is the second aspect of our method which we would like to stress. It seems to us that copying students would be unlikely to go to the trouble of disguising universal syntactic markers.

Acknowledgments: The work of Nicholas Heard described in this paper was funded by the Wellcome Trust, grant number 065822.

Appendix

6   Example of a Coursework Question and Solution

Question:

1. For each of the functions, f, below, find the values of c which make the functions legitimate probability density functions.

   a. $f(x) = \begin{cases} c, & a \le x \le b \\ 0, & \text{otherwise} \end{cases}$

   b. $f(x) = \begin{cases} ce^{-\lambda x}, & x > 0 \\ 0, & \text{otherwise} \end{cases}$

   c. $f(x) = c\exp\left[-\frac{(x-\mu)^2}{2\sigma^2}\right]$

2. What are the mean values of the distributions given in part (i)?
3. What are the median values of the distributions given in part (i)?
4. A user of the internet normally uses search engine A, but is thinking of switching to search engine B.

   a. It is known that the times taken to locate particular items of information on the internet vary from search to search, and that the distribution of these times is right skewed. Draw a sketch indicating the shape of such a distribution

   b. The log transforms of the search times are known to follow a normal distribution fairly closely. The mean log(time) for search engine A to locate items is known to be 1.5. A user is considering switching to search engine B, and has collected the information below, which shows the log(time) values for 10 randomly chosen searches. Compute the mean and standard deviation of the log(times) in the sample.

   2.6 2.2 1.5 1.4 1.4 1.2 1.8 1.1 1.0 2.9

    c. Using appropriate tables from the formula sheet, carry out a test of the hypothesis that the log(time) values using engine B are drawn from a distribution with a mean of 1.5, at the 5% level. In your answer, clearly state which distribution you use for the test statistic, and write down any formulae you use to compute the test statistic.

    d. What recommendation would you make to the user?

Solution:

1. a. This is a uniform distribution, with constant height. Since $\int f(x)\,dx = 1$, we must have $1 = \int_a^b c\,dx = \frac{c}{b-a}$, so that $c = 1/(b-a)$.

    b. We must have $1 = \int_0^\infty ce^{-\lambda x}\,dx = c\int_0^\infty e^{-\lambda x}\,dx = c\left[e^{-\lambda x}/(-\lambda)\right]_0^\infty = \frac{c}{\lambda}$, so that $c = \lambda$.

    c. $c = 1/(\sigma\sqrt{2*\pi})$ by recognising that the distribution is normal.

2. a. mean $= (b+a)/2$ either by a symmetry argument, or by integration

    b. mean $= 1/\lambda$ either from the formula sheet, or by integration

    c. mean $= \mu$ from the formula sheet or memory.

3. a. median $= (b+a)/2$ by a symmetry argument

    b. median $= \frac{\log 2}{\lambda}$ found from $\int_0^m \lambda e^{-\lambda x} dx = \frac{1}{2}$, by integration, or from $1 - e^{-\lambda x} = 1/2$ from the formula sheet.

    c. median $= \mu$, by symmetry

4. a. Any sketch of a right-skewed distribution will do.

    b. $\sum x = 17.1 \quad \bar{x} = 1.71$

$$\sum x^2 = 33.07$$

$$Var = \frac{\sum x^2}{n-1} - \frac{n\bar{x}^2}{n-1} = \frac{33.07}{9} - \frac{10 \times 1.71^2}{9} = 0.425$$

So sd $= 0.652$

OR $\sum (x - \bar{x})^2 =$
0.7921+0.2401+0.0441+0.0961+0.0961+
0.2601+0.0081+0.3721+0.5041+1.4161
$= 3.829$

So that sd $= \sqrt{\sum (x - \bar{x})^2 / 9} = \sqrt{3.829/9} = 0.652$

    c.

$$t = \frac{\bar{x} - \mu}{s/\sqrt{n}} = \frac{1.71 - 1.5}{0.652/\sqrt{10}} = \frac{0.21}{0.206} = 1.02$$

From the t-tables in the formula sheet, referring to the row for 10-1 = 9 degrees of freedom, we see that this is less than 2.26 and hence is not significant at the 5% level. We have no reason for supposing that search engine B has log(time) values different from that for search engine A

    d. This test provides no reason to change search engines.

7 Extract from the Data Matrix

Rows represent students (in fact there were 50) and columns represent the counts
of +, -, and = in each part of each question, yielding 60 columns in all.

12	15	1	1	2	4	0	8
12	5	1	1	6	4	0	10
9	13	4	1	1	3	0	8
12	15	1	1	5	4	0	8
2	4	2	0	8	4	0	10
2	2	1	0	3	2	0	3
10	14	0	2	6	0	0	0
1	3	0	0	0	0	0	0
0	0	0	0	0	0	0	0
9	12	6	0	7	2	0	3
7	11	2	0	10	2	0	4
16	19	2	0	10	4	0	10
9	5	4	0	7	6	0	12
24	6	0	0	0	0	0	0
7	5	1	0	4	4	0	1

References

[1] Hand D.J., Adams N.M., Bolton R.J. (eds.): *Pattern Detection and Discovery.*
(2002) Springer
[2] Hand D.J., Bolton R.J. : Pattern discovery and detection. *Journal of Applied Statistics,* (2004) to appear
[3] Truss L. : *Eats, Shoots, and Leaves: the Zero Tolerance Approach to Punctuation.*
(2003) London: Hatton Books
[4] Glaz J., Naus J., Wallenstein S.: *Scan Statistics.* (2001) New York: Springer

Local Pattern Detection and Clustering
Are There Substantive Differences?

Frank Höppner

University of Applied Sciences Braunschweig/Wolfenbüttel
Robert Koch Platz 10-14
D-38440 Wolfsburg, Germany

Abstract. The starting point of this work is the definition of local pattern detection given in [10] as the unsupervised detection of local regions with anomalously high data density, which represent real underlying phenomena. We discuss some aspects of this definition and examine the differences between clustering and pattern detection (if any), before we investigate how to utilize clustering algorithms for pattern detection. A modification of an existing clustering algorithm is proposed to identify local patterns that are flagged as being significant according to a statistical test.

1 Introduction

Knowledge discovery in databases (KDD) aims at detecting valid, novel, potentially useful, and ultimately understandable patterns in data [8]. Many tools in KDD aim at a global characterization of the data, such as decision trees or clustering partitions. The more recent technique of association rule mining, however, investigates into more local phenomena that do not characterize the database as a whole but only a small subpopulation. Usually, association rule mining is considered as the most prominent approach to *local pattern detection*. However, experiments with (standard) association rule mining are often somewhat frustrating, because the number of local patterns often becomes that large that it is no longer manageable. And even worse, most of these patterns – flagged as being potentially interesting – turn out to be neither useful nor valid in the application context. For a deeper discussion see [4]. The definition of local pattern detection given by Hand [10] takes these aspects into account. The main points in his definition are:

1. A local pattern is a data vector serving to describe an *anomalously high local density of data points* when compared to a background model:

$$\text{data} = \text{background\_model} + \text{pattern} + \text{random\_component} \qquad (1)$$

2. Local pattern detection is unsupervised in the sense that no information but the data itself is given to find out what patterns may be present in the database, if any.

K. Morik et al. (Eds.): Local Pattern Detection, LNAI 3539, pp. 53–70, 2005.

3. Local pattern detection is about inferring from observations, therefore patterns must represent real phenomena and not just noise.

In this paper we will contrast the goals of local pattern detection with those of clustering (section 2) and discuss some potential problems and consequences when following the definition above (section 3). Whether a flagged pattern is substantive or not is influenced by two different facts, one is the statistical significance of an identified candidate patterns, the other is the robustness of the applied algorithms, that is, the sensitivity to initial parameters, which is a problem with many clustering algorithms in particular. We will discuss consequences and candidate algorithms in section 4. In section 5 we will finally discuss a pattern detection algorithm that has many of the desired properties discussed before, which will be illustrated via some examples in section 6.

2   Local Pattern Detection Vs. Clustering

At first glance, the before-mentioned description of pattern detection sounds almost identical to clustering. Here is an exemplary definition from the literature:

"Clusters may be described as connected regions of multi-dimensional space containing a relatively high density of points, separated from other such regions by a region containing a relatively low density of points" [7]

The identification of (local) regions with high data density (point 1 in the definition) and the fact that pattern detection is an unsupervised approach (point 2) establishes a strong relationship between local pattern detection and clustering.

In accordance with point 1 of the definition, we could compose our data model out of several Gaussian distributions and a single uniform distribution. If we think of the uniform distribution as the background model in (1) and the Gaussian distributions as the patterns, the differences between pattern detection and clustering begin to blur. Standard mixture decomposition could be applied to identify the parameters of the models – and if the parameters of the Gaussian indicate that only a small portion of the input space is affected (small covariance), we could speak of an identified pattern.

May be it is surprising that traditional definitions of clustering [7, 12, 14] do not contain anything similar to the third statement in the definition of local pattern detection, which refers to the statistical "validity" of identified clusters.[1] While it is not mentioned in the definitions, the problem that "the resulting clustering procedures have no known significant theoretical properties" [5] is well recognized. But unfortunately not much has changed since Hartigan stated in 1975 [12] that clustering algorithms "are not yet an accepted inhabitant of the

---

[1] With some clustering algorithms, e.g. when the number of clusters has to be fixed in advance, so called *validity measures* are used to "validate" the results. Even if these measures are not purely heuristic in nature but investigate statistical properties of a partition, they seldomly take the role of a statistical test.

statistical world". This makes the current position in pattern detection even more similar to that in clustering, because in both fields some theoretical framework is missing. (Given that Hartigan made his statement in 1975 and also given the lack of progress in this concern, the "development of a theoretical base" [10] for pattern detection appears really challenging.)

Rather than by using statistical tests, in machine learning overfitting is often avoided by employing a regularization framework. In contrast to statistics, such a framework aims at limiting the variability of the models, but does not care primarily about the statistical significance of the result. On the other hand, if the assumptions of the used statistical model (which are always present) are violated (which may happen quite easily in KDD) there is not much left that distinguishes regularization from statistical relevance tests.

Up to this point the reader may agree that clustering and (today's) local pattern detection are indeed very similar. The only distinction that is left is the explicit focus on *local* patterns, which cannot be found in clustering. We will see in the following, however, that this is not enough for a substantive distinction. So a provocative definition of local pattern detection could be "clustering, done right".

3   What Is the Background Model?

Having a background model defining the normal situation enables us to apply a statistical test to see whether some observations deviate significantly from the background model or not. Thus the background model plays a key role in detecting substantive local patterns. On closer inspection, however, it becomes clear that this works well only under the assumption that the background model is valid. And determining the validity of the background model may be as difficult as determining the validity of a cluster (or pattern) without a supporting background model.

This leads us to the general question of how to select a background model. A good candidate for a background model, when little domain knowledge is available, might be the uniform distribution. Figure 1(a) illustrates a hypothetical data set. On the right hand side the data density is 4.0 (per some area) and on the left it is 2.0, both sides are occupied by approximately the same number of data objects. On both sides, there are smaller regions in which the data density is 3.0; intuitively, these are the "patterns". If we assume a uniform distribution as the background model, we obtain an average density of 3.0, which perfectly corresponds to the density of our patterns. Therefore, *this* background model would not flag them as substantive patterns. The background model may flag the larger regions as deviations from the background model, but they do not qualify as patterns due to their size. (By the way, does the small cluster on the right qualify as a pattern? It represents a deviation from the background model, but its data density is smaller rather than larger.)

The point in Fig. 1 is of course that a single, simple background model will not work. Either the background model must be flexible and complicated, or it

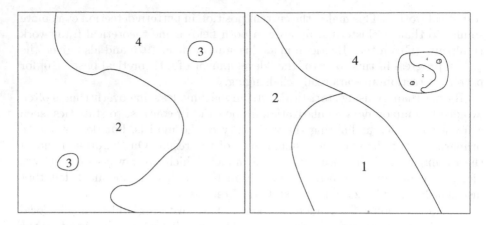

(a) The numbers in the areas denote the average data density per volume.

(b) Locality is a matter of scale: The environment of figure.

Fig. 1. A hypothetical data set.

must be possible to define different background models in different parts of the data space.[2] In the context of clustering, we could say that we have two clusters in Fig. 1(a), the left and the right part of the figure. And within each cluster, there is a small subcluster – which we may call a local pattern. But indeed, we never know whether we currently observe a cluster, a background model or a local pattern, unless we know about the scale at which we look on the data. (The whole Fig. 1(a) may be a local pattern itself – in the upper right corner of the coarser view of figure 1(b).) Even though we may be interested in local patterns only, we *have to* carefully consider structures at any larger scale. In analogy to (1) we could try to express this fact by a recursive definition

$$\text{data} = \text{model} + \text{noise} \qquad \text{where}$$
$$\text{model} = \text{atomic\_model} \mid \text{background\_model} + \text{model}^* \qquad (2)$$

where model* means that any number of models can supplement the background model. Thus, the data may be represented by a hierarchical tree of models, where the same model may serve as a cluster in one level and a background model in another. Local patterns (in the sense of small clusters) can be considered as the leaves of this model tree.

A background model helps with the identification of substantive local patterns only if the background model itself is valid. Simple examples (fig. 1(a)) show

---

[2] In association rule mining, the minimum support threshold may be seen as being part of a very coarse background model. In a k-dimensional boolean space, we have 2^k possible configurations. For n records and a uniform distribution, we expect $\frac{n}{2^k}$ objects per combination. The $\min_{supp}$ threshold, however, is the same for *all item sets of any size* and does not depend on k.

that we cannot restrict ourselves to a global, simple background model. Estimating a valid background model of arbitrary complexity (eq. (1)) in one step seems unrealistic. Utilizing erroneous or inadequate background models puts the validity of the identified local patterns in question. The most promising approach is to start with a simple model (whose parameters can be estimated easily and robustly) and use this as the basis for the next hierarchy level to come (eq. 2)). Then, for the identification of all models in later stages, we already benefit from the existence of a valid background model (stepwise refinement). This approach allows us to stick to simple background models, such as the uniform distribution, even in cases like figure 1(a) (only the boundary remains to be determined). And this approach also underlines that we cannot focus on local structure only, but must carefully investigate structures at any scale.

4   Escaping from Heuristic Thresholds

One can easily find clustering algorithms that respect these ideas. For instance, we could start by estimating the parameters of a mixture of Gaussians. In [5] (p. 558) a statistical test is proposed to decide whether a cluster may have been generated from a single Gaussian. Such tests can be applied to each cluster for validation; if the test fails, we may generate a new data set that contains only the data objects belonging to this cluster (for instance, we could sample from the original data set using the a posteriori probabilities of being generated by the Gaussian). We then apply the same clustering algorithm once more, which leads us to a hierarchical subdivision of the previously discovered cluster. The data set refinement is stopped if such a refinement cannot be justified by the data any longer. A similar method (where the tests are not statistical in nature) can be found in [9].

When implementing such algorithms technical details become highly relevant for failure or success, such as:

- Did the algorithm yield the correct solution (that is, did the expectation maximization algorithm yield the (globally) optimal solution)?
- Was the assumption of Gaussian distributions justified?
- If we need data density estimations (e.g. to detect the clusters in Fig. 1(a)), did we select an appropriate size of the area that is used to estimate the density?

Most clustering algorithms require a couple of initialization parameters – and are generally more sensitive to their setting than we would like them to be. The (background) models are not the only information we are processing, and the same effort to validate background models and patterns should also be spent on any other step in the line of processing, because invalid intermediate results also deteriorate the correctness of our final patterns. The theoretical advantage of using statistical tests with the background model is worth nothing if the algorithms pass ill-formed pattern candidates to the test.

With every (heuristic) threshold an algorithm requires we increase the risk of processing unvalidated data. And a lot of decisions may be necessary, in particular in the preprocessing phase. Most often, the parameters are chosen on the basis of some small sample and visual inspection, but in KDD we cannot be sure that the parameter will be valid for all unseen data to come. In fact, there might be no single parameter that suits all local patterns equally well.

In the recent past, the *multiscale* approach has turned out to be a powerful weapon against this problem: Rather than choosing one parameter setting, examine the results for *all possible settings* and choose the single or multiple values that yield the *most stable* results. Multiscale techniques have been proven extremely helpful in many areas, such as image and shape recognition [15], signal analysis [13, 16], data compression [17], and also clustering [2, 1], to mention only a few. The next section briefly summarizes the OPTICS algorithm [1], a multiscale clustering algorithm, which will be used in the subsequent sections for pattern detection purposes.

Multiscale Clustering

In this section we will informally introduce the OPTICS algorithm, for the full details we refer to [1]. Density based clustering algorithms usually count all data objects within a hypersphere (or hyperbox) of fixed size to obtain a density estimate. We say the neighborhood $N_\varepsilon(q) = \{x \mid \|x - q\| \leq \varepsilon\}$ of a point q in the database D is dense, if $|N_\varepsilon(q)| \geq k$. Given k and ε, a cluster C is defined as a non-empty set, which satisfies two conditions: (a) a cluster has at least one point with a dense neighborhood and (b) for each point $p \in C$ with a dense neighborhood, $N_\varepsilon(p) \subseteq C$ holds. Since the identified clusters depend on the choice of ε, we speak of ε-clusters. The DB-SCAN algorithm [6] determines all clusters (with respect to ε and k) in $O(n \log n)$ where n is the number of points.

The choice of ε is crucial in the DB-SCAN algorithm, and often it is not possible to discover all the structure in a dataset with a single choice of ε. The idea of the OPTICS algorithm is to generate all partitions for all possible values of ε within some range $[0, \varepsilon_{\max}]$ (in an efficient way). But then it remains still unclear how to interpret or analyze that many resulting partitions. An interesting question to ask is at what distance ε a point p's neighborhood will become dense (called core distance) and at what distance a point p will belong to a cluster for the first time (called reachability distance). (Apparently the reachability distance is less than or equal to the core distance, because at the core distance the point will become a cluster of its own.) The OPTICS algorithm determines these two values for all data objects and, furthermore, an ordering of data objects that allows for a reconstruction of any DB-SCAN partition (see [1]). Figure 2(a) shows an example of the so-called reachability plot, which aligns the data objects according to the determined ordering on the horizontal axis. For any point p in the plot (e.g. the marked one in Fig. 2(a)), the data points with smaller reachability values to the left make up a (DB-SCAN-) cluster at the chosen value of ε.

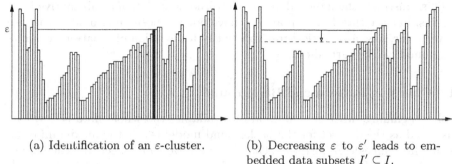

(a) Identification of an ε-cluster.

(b) Decreasing ε to ε' leads to embedded data subsets $I' \subseteq I$.

Fig. 2. The reachability plot (result of the OPTICS algorithm). Horizontal axis: point ordering, vertical axis: reachability value

Now it should be clear, how clusters (and local pattern candidates) are found in the reachability plot: Clusters are "dents" (or valleys) in the graph, indicating a region of high data density surrounded by data with lower density. Since the width of a valley is determined by the number of data objects in the cluster, we can use the width to distinguish large from small clusters (patterns).

5 An Approach to Local Pattern Detection

In [1] a heuristic procedure is proposed to extract clusters automatically from the reachability plot. Thresholds on the steepness and length of the flanks surrounding a flat valley are used to identify clusters. Although this technique seems to work well, a drawback is the need for selecting a new heuristic parameter.

Here, we choose a different approach. Two things are needed in order to detect substantive patterns: the pattern itself and the background. For the moment we are not concerned about *what* model we will actually use, but about the data subset that will be used to estimate the model's parameters (pattern as well as background model). A reasonable way to identify subsets is to consider all data objects that are density connected for some ε (that is, belong to the same ε-cluster). Local regions of high data density can be obtained from the reachability plot by drawing a horizontal line at ε_P. Each interval on the data axis, where the reachability plot drops below this line, corresponds to a data subset in which all points are density-reachable at ε_P. Let us denote the data objects associated with such an interval by I_P and denote the number of points by n_P. When decreasing ε_P the subsets become more dense and smaller (cf. Fig. 2(b)).

Since we need two subsets, a larger one that corresponds to background and a smaller one that corresponds to the pattern, we simply draw another horizontal line at some larger $\varepsilon_B > \varepsilon_P$. For each local pattern subset I_P we obtain a background subset I_B with $I_P \subseteq I_B$. Now, if the pattern model P (estimated from data in I_P) deviates from the background model B (estimated from data in I_B) significantly, we have identified a substantive pattern.

This illustrates the intended approach to the detection of a substantive local patterns, but the thresholds ε_P and ε_B have not yet been determined. It is also not yet clear, how a statistical test to identify a deviation of a pattern from its background can be carried out.

5.1 Choosing Pattern and Background

At the beginning, with not information available, the whole data set will be considered as the dataset for the background model (ε_B is the maximum of all reachability values). From the reachability plot we can collect all reachability values that actually occur and scan them from the largest (current background) to the smallest value. For every new value ε we pass, we have one or more data objects whose reachability value is identical to ε. Since large reachability values indicate that there is a larger gap between the data to the left and on the right, such a data point subdivides the current data subset into two or more parts (cf. Fig. 3(a)). If a statistical test (that still has to be developed) indicates that there is a significant deviation of one of the these subset from the current background, we mark this subset as a cluster (or deviation from the background). If we move the scan line further downwards, this new subset serves itself as the new background model for subsequent subdivisions, as illustrated in figure 3(b). In this way, we create a hierarchical tree of subsets directly from the reachability graph, similar to the one discussed in [13].

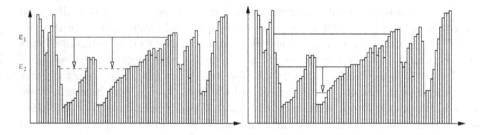

(a) If ε_1 decreases, the associated data subset is split up by a peak in the reachability plot.

(b) Once a significant deviation has been identified, the newly identified pattern plays the role of the background as ε decreases further.

Fig. 3. Identification of background and pattern.

5.2 A Pattern Test

In the following we need local data density estimates. To calculate the data density we need to approximate the *space* that is occupied by a subset of the data. To get this estimate, we use the second outcome of the OPTICS algorithm, the core density of each data point. This is the distance to the kth neighbor and can therefore be used for local data density estimation[3]. Given that for a data object x the distance to the kth neighbor in the d-dimensional space is r, on average it occupies the space $V_x = \frac{V}{k}$, where $V = \frac{\sqrt{\pi^d}}{\Gamma(d/2)} r^d$ is the space occupied by the sphere containing the k nearest neighbors of x and Γ denotes the Gamma function. In the two-dimensional case of our illustrative examples, we assign to each data object x a volume of $V_x = \frac{\pi \cdot r^2}{k}$. The volume that is occupied by a subset of the dataset is simply the sum of volumes of each data point within the pattern or background. It should be noted that this estimation contains only the *occupied* space and free space in between is not considered. For instance, if we have two uniform clusters of identical density, the estimated volume for this data set contains the volume of the clusters only, but not the space between the clusters.[4]

There are several possibilities for defining models for patterns and background. For instance, we could use a uniform data density; we may assume that the data objects are uniformly distributed in the occupied data space, and that we have found a substantive cluster if for some subset the number of data objects differs significantly from the expected number of data objects given the volume of this subset. Having assigned data volumes V_P to the pattern and V_B to the background, we can define a binomial distribution where the probability of a randomly chosen data object lying in the pattern volume is simply $p = \frac{V_P}{V_B}$. The expected number of data objects in the pattern is then $n_B \cdot p$, which can be tested against the actual number of data objects n_P (where n_P is the number of data objects in the pattern and n_B in the background). Unfortunately, this approach fails in practice. Suppose we have a data set generated completely at random from a uniform distribution. It may happen that a few data points, say 3, are by chance very close together, much closer than the average distance between data objects. This leads to a very small total volume for this subset. Any background set occupies much larger space V_B, which leads to very small pattern probabilities p. Such small probabilities make the chances of generating 3 data objects within the pattern region very unlikely even for small background sample sizes. In consequence, this approach flags much more patterns as being significant than there are actually in the dataset.

It is also possible to assume that the local data densities within a subset obey some known distribution and to test the parameters obtained from the pattern and the background for being identical. But from the construction of

---

[3] We used a value of 5 for k to limit the influence of border effects. Larger values are better for visual inspection of the reachability plot, but if a pattern consists of a few points only and k is high, it is very likely that the density estimation is heavily influenced by the surrounding data that do not belong to the pattern whose density we want to estimate.

[4] This is quite different from those approaches to clustering where assumptions on certain cluster shapes are made, such as hyperspherical clusters with k-means and derivatives. There, cluster volume estimations are usually based on the center and some mean distance between data objects and center.

the subsets via the reachability plot it is clear that the pattern sample is not a random sample of the background subset, but we intentionally consider only those data values that have a small data volume. Therefore it is quite obvious that we will observe significant deviations in, say, the mean density of pattern and background quite frequently.

The approach that evaluated best is the following: Let ϱ_i be the data density estimated for data object x_i and N be the number of data objects, $\varrho_{min} = \min\{\varrho_i | 1 \leq i \leq N\}$ and $\varrho_{min} = \max\{\varrho_i | 1 \leq i \leq N\}$. The range $[\varrho_{min}, \varrho_{max}]$ of estimated data densities is partitioned into m equally sized parts

$$S_i = [\varrho_{min} + (i-1)\Delta, \varrho_{min} + i \cdot \Delta]$$

with $\Delta = |\varrho_{max} - \varrho_{min}|/m$ (in the experiments m was set to 24). We consider the local data density as being an attribute of the data object itself rather than a property of its neighborhood. Thus B (resp. P) is a m-nomial random variable whose outcome determines the density of a point in the background (resp. pattern) dataset; $P(B = S_1)$ denotes the probability of a randomly chosen data object to 'have' a data density within $[\varrho_{min}, \varrho_{min} + \Delta]$. The distribution $P(B = S_i)$ is empirically estimated from $|\{x_j | \varrho_{min} + (i-1)\Delta \leq \varrho_j < \varrho_{min} + i\Delta\}|/N$.

A chi-square test can be applied to test whether a sample (the pattern subset) may have been generated from this multinomial distribution. In this case, the pattern would not be flagged as a deviation from the background. But before we apply this test, we compensate for the subset selection bias mentioned in the previous paragraph. The deeper the subset is in the hierarchy (or the smaller ε_P is), the higher the data density will be. We therefore do not compare the m-nomial distributions, but exclude the part of $\mathbf{Pr}(B)$ with low data densities, which are no longer present in the subset due to the way we select the subset from the reachability graph. That is, we find a lower bound ϱ for the density values in the subset and compare $\mathbf{Pr}(B|B > \varrho)$ with $\mathbf{Pr}(P|P > \varrho)$ rather than $\mathbf{Pr}(B)$ with $\mathbf{Pr}(P)$. As an example, assume the background data density distribution is given by

$$(0.0, ..., 0.0, 0.01, 0.0, 0.03, 0.05, 0.07, 0.10, 0.09, 0.13, 0.21, 0.12, 0.11, 0.08)$$

that is $P(B = S_m) = 0.08$, $P(B = S_{m-1}) = 0.11$, etc. Starting from the left (S_0, sparse data, low data density), we calculate the number of data objects that we expect in the pattern subset with this data density, given the size $|P|$ of the current pattern candidate P. If this expected number is below 5 or no data objects with this data density were observed in the pattern subset, the chi-square test cannot be applied and we consider a reduced $(m-1)$-nomial distribution with the leftmost slot removed. This step is repeated and the number of slots is reduced to some $0 \leq m' \leq m$. In the example, for $|P| = 100$, $m' = 9$. With $\varrho = \varrho_{min} + m' \cdot \Delta$, the distribution $P(B|B > \varrho)$ (that is, only a m'-nomial distribution) is then tested against $P(P|P > \varrho)$. This procedure is to some degree a technical necessity to apply the chi-square test, but also effectively excludes regions of low data density in the background in the comparison with the pattern candidate and thereby compensates the discussed pattern selection bias.

Figure	number of data objects			
	noise	pattern 1	pattern 2	pattern 3
5	2000	–	–	–
4(a)	2000	50	–	–
4(b)	1500	250	250	100
4(c)	1500	400	100	100
4(d)	2000	50	30	20

Figure	mean values		
	pattern 1	pattern 2	pattern 3
4(a)	$\begin{pmatrix} 0.3 \\ 0.3 \end{pmatrix}$	–	–
4(b)	$\begin{pmatrix} 0.3 \\ 0.7 \end{pmatrix}$	$\begin{pmatrix} 0.7 \\ 0.2 \end{pmatrix}$	$\begin{pmatrix} 0.8 \\ 0.7 \end{pmatrix}$
4(c)	$\begin{pmatrix} 0.4 \\ 0.5 \end{pmatrix}$	$\begin{pmatrix} 0.7 \\ 0.2 \end{pmatrix}$	$\begin{pmatrix} 0.8 \\ 0.7 \end{pmatrix}$
4(d)	$\begin{pmatrix} 0.4 \\ 0.6 \end{pmatrix}$	$\begin{pmatrix} 0.2 \\ 0.4 \end{pmatrix}$	$\begin{pmatrix} 0.9 \\ 0.1 \end{pmatrix}$

Figure	covariances		
	pattern 1	pattern 2	pattern 3
4(a)	$\begin{pmatrix} 0.025^2 & 0 \\ 0 & 0.025^2 \end{pmatrix}$	–	–
4(b)	$\begin{pmatrix} 0.1^2 & 0 \\ 0 & 0.05^2 \end{pmatrix}$	$\begin{pmatrix} 0.1^2 & 0 \\ 0 & 0.1^2 \end{pmatrix}$	$\begin{pmatrix} 0.05^2 & 0 \\ 0 & 0.05^2 \end{pmatrix}$
4(c)	$\begin{pmatrix} 0.2^2 & 0 \\ 0 & 0.2^2 \end{pmatrix}$	$\begin{pmatrix} 0.05^2 & 0 \\ 0 & 0.1^2 \end{pmatrix}$	$\begin{pmatrix} 0.05^2 & 0 \\ 0 & 0.05^2 \end{pmatrix}$
4(d)	$\begin{pmatrix} 0.05^2 & 0 \\ 0 & 0.05^2 \end{pmatrix}$	$\begin{pmatrix} 0.02^2 & 0 \\ 0 & 0.03^2 \end{pmatrix}$	$\begin{pmatrix} 0.02^2 & 0 \\ 0 & 0.02^2 \end{pmatrix}$

Table 1. Construction of the data sets in figure 4 (number of global noise points, mean and covariances of local patterns).

6   Examples

In this section we present some results obtained from the proposed local pattern detection algorithm. We discuss results for five data sets, one of them consisting of 2000 data objects uniformly distributed in the unit square. All other data sets are depicted in figure 4(a)-4(d). The dataset in Fig. 4(a) has also been used in [3]. Table 1 summarizes how the data sets have been generated. Especially Fig. 4(d) represents a difficult problem, because the superimposed patterns are really small and quite difficult to identify even for a human.

Figure 5 shows the reachability graph for the uniform data set and two different values of k (number of data points in a dense neighborhood). Although no substantive patterns were superimposed over the uniform noise, the reachability

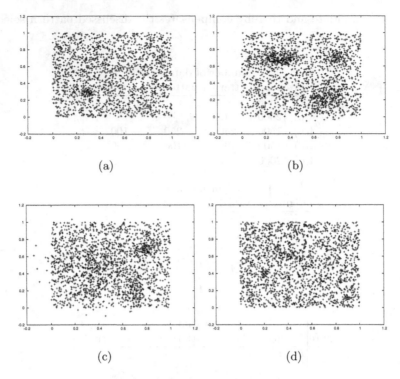

Fig. 4. Collection of test data sets, generated according to table 1.

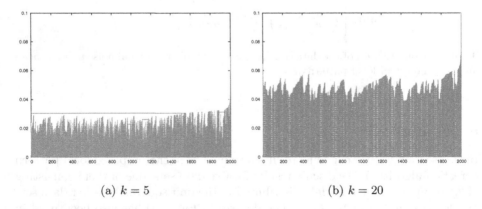

Fig. 5. Reachability plot for uniform distribution with $k = 5$ and $k = 20$.

plot shows many random local minima and maxima, which are more distinct for $k = 20$. For all experiments $k = 5$ has been used because 20 data points is already the size of the smallest pattern we want to discover in Fig. 4(d) (cf. also footnote 3).

Figure 5(a) additionally shows via horizontal lines the data subsets that were identified as substantive patterns by the algorithm, as it was discussed in Fig. 3. Four such intervals have been determined, one contains almost 75% of the data set and therefore would not qualify as a small cluster or pattern. Given the number of flagged patterns reported in [3], showing only 4 substantive pattern/background-deviations (only 3 qualify as potential patterns) is an impressive small number. The four identified subsets are shown in Fig. 6. The top left figure corresponds to the long line, the top right figure to the short line to the right. The two figures in the bottom correspond to the small patterns that use the "long line" subset as the background pattern. In both of these subpatterns the data density deviates by chance significantly from the data density in the background.

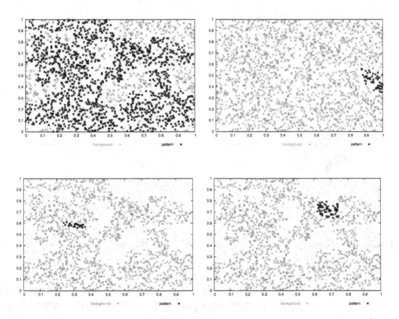

Fig. 6. Flagged clusters in the uniform data set. Top row: the whole dataset is the background data, the black points are the pattern. The pattern in the top left figure corresponds to the long line in Fig. 5(a). Bottom row: The identified pattern in the top left figure became the background (gray) for the two patterns in the bottom row. The patterns are again shown in black, the background in gray. The points in light gray do not belong to background nor pattern.

For the dataset in Fig. 4(a) with a single substantive cluster, five pattern / background combinations have been identified. They are depicted in the reachability graph in figure 7. Four of the five subsets are subsets of each other (the algorithm focuses slowly on the core of the pattern), such that only the "smallest" subset qualifies as a local pattern. Two of these hierarchically embedded subsets are shown in the bottom row of images in Fig. 7, with the smallest cluster (right bottom) corresponding very well to the superimposed normal distribution. The single remaining subset is shown in the top right image, which identifies another region of particularly high data density. This pattern is not artifically generated but occurred by chance, but *only one* such incidental agglomeration has been flagged.

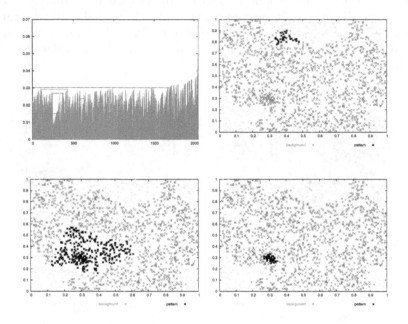

Fig. 7. Flagged clusters in the data set of Fig. 4(a). (See also explanations in Fig. 6.)

The results of the datasets in Fig. 4(b) and 4(c) are shown in Fig. 8 and 9, resp. In both cases we have quite large patterns, but different data densities. The data densities of the patterns in Fig. 4(b) deviate clearly from the background noise. Similar to the previous case, the algorithm determines a sequence of significant deviations that slowly focuses on a small spot, which can then be considered as a local pattern. Although the number of marked subsets is quite large in Fig. 8 (top left), we have only five different local patterns identified. The three largest correspond to the superimposed patterns and are shown in the figure. The fact that – compared to Fig. 8 – much more deviations have been recognized is due to the fact that Gaussian distributions have been superimposed: rather than an abrupt change in the density, which would lead to a single deviation, we have a slowly increasing data density which introduces several significant deviation levels. If we are interested in local patterns only, we can ignore all those patterns that contain an even smaller subpattern, which leads us again to a very small number of flagged local patterns.

In contrast to Fig. 4(b), the data densities of the patterns in Fig. 4(c) do not deviate that much from the background density, but this does really affect the performance of the algorithm, as we can see from Fig. 9. We have fewer focusing steps, but again the smallest patterns correspond to the superimposed Gaussian distributions. Besides the three true patterns, only one more false positive pattern has been flagged.

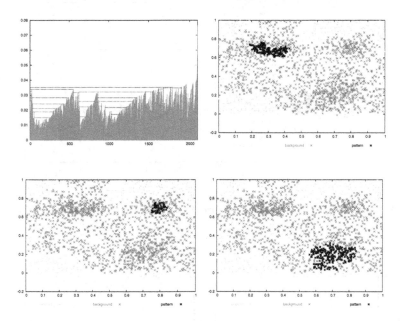

Fig. 8. Some flagged clusters in the figure 4(b). (See also explanations in Fig. 6.)

Finally, Fig. 10 shows the results for the most difficult test set in Fig. 4(d). Five local patterns are identified, three of them correspond to the true patterns, we have only two false positives.

7   Summary and Conclusions

We have seen that local pattern detection (in the notion of [10]) is very similar to clustering. The challenge in local pattern detection is almost the same as in clustering, namely to identify valid, substantive structure (clusters, patterns) in data. The smaller the structure, the more difficult it is to determine its validity, because smaller structures are more likely to occur by chance.

To tackle this problem, it was proposed in [10] to install a (global) background model to verify local patterns against the background. The feasibility of the approach depends on the validity of the background model, but we have seen

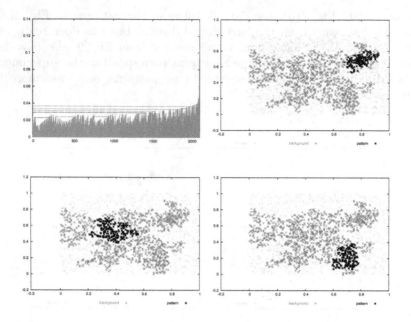

Fig. 9. Some flagged clusters in the figure 4(c). (See also explanations in Fig. 6.)

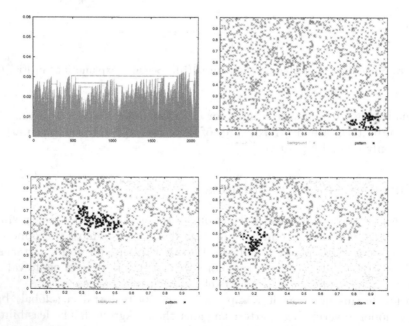

Fig. 10. Some flagged clusters in the figure 4(d). (See also explanations in Fig. 6.)

that we cannot restrict ourselves to simple background models. Therefore, a hierarchical approach appears to be most promising: instead of estimating a global complex background model, the utilization of a tree of simple (background) models is proposed, where each of them is installed only if it significantly deviates from the previous model. The hierarchical approach underlines that local pattern detection cannot be concerned about the local structures *only*, but has to carefully investigate structures at any scale – just like clustering.

Flagging a pattern candidate as being substantive or not is one thing, but the same care should be applied to the identification of pattern candidates (which are then passed to the statistical test). The more heuristic parameters an algorithm utilizes, the higher the chances of choosing inappropriate values. If the results are very sensitive to these parameters, how can we be sure that we identify real patterns or just artefacts? Multiscale algorithms have the advantage that they do not count on the user's 'guessing' capabilities but almost eliminates a threshold by analyzing the results over a large range of possible settings.

The OPTICS algorithm is a clustering algorithm that satisfies most of these requirements: it is a multiscale algorithm, is quite insensitive to the choice of the only parameter k, detects clusters of arbitrary shape and is efficient. We have discussed an alternative 'backend' to this algorithm that identifies a tree of significant deviations, whose leaves correspond to local patterns. For a number of two-dimensional test cases the results were shown: all patterns have been identified and only a very small number of false positives have been flagged. Validating the approach in a broader set of test data remains for future work.

Acknowledgments: Many thanks to Prof. Dr. Kriegel for kindly providing an implementation of the OPTICS algorithm.

References

[1] Mihael Ankerst, Markus M. Breunig, Hans-Peter Kriegel, and Jörg Sander. OP-TICS: Ordering points to identify the clustering structure. In *Proc. of ACM SIGMOD Int. Conf. on Management of Data*, Philadelpha, 1999.

[2] Gerardo Beni and Xiaomin Liu. A least biased fuzzy clustering method. *IEEE Trans. on Pattern Analysis and Machine Intelligence*, 16(9):954–960, September 1994.

[3] Richard J. Bolton and David J. Hand. Significance tests for patterns in continuous data. In *Proc. of IEEE Int. Conf. on Data Mining*, 2001.

[4] Richard J. Bolton, David J. Hand, and Niall M. Adams. Determining hit rate in pattern search. In *[11]*, pages 36–48, 2002.

[5] Richard O. Duda, Peter E. Hart, and David G. Stork. *Pattern Classification*. John Wiley & Sons, 2001.

[6] Martin Ester, Hans-Peter Kriegel, Jörg Sander, and Xu Xiaowei. A density-based algorithm for discovering clusters in large spatial databases with noise. In *Proc. of the 2nd ACM SIGKDD Int. Conf. on Knowl. Discovery and Data Mining*, pages 226–331, Portland, Oregon, 1996.

[7] B. S. Everitt. *Cluster Analysis*. John Wiley & Sons, 1974.

[8] Usama M. Fayyad, Gregory Piatetsky-Shapiro, Padhraic Smyth, and Ramasamy Uthurusamy, editors. *Advances in Knowledge Discovery and Data Mining*. MIT Press, 1996.

[9] Amir B. Geva. Non-stationary time-series prediction using fuzzy clustering. In Rajesh N. Davé and Thomas Sudkamp, editors, *Proc. of the 18th Int. Conf. of the North American Fuzzy Information Processing Society*, pages 413–417, June 1999.

[10] David Hand. Pattern detection and discovery. In *[11]*, pages 1–12, 2002.

[11] David Hand, Niall M. Adams, and Richard J. Bolton, editors. *Pattern Detection and Discovery*, volume 2447 of *LNAI*. Springer, 2002.

[12] John A. Hartigan. *Clustering Algorithms*. John Wiley & Sons, 1975.

[13] Frank Höppner. Handling feature ambiguity in knowledge discovery from time series. In *Proc. of 5th Int. Conf. on Discovery Science*, number 2534 in LNCS, pages 398–405, Lübeck, Germany, November 2002. Springer.

[14] Anil K. Jain and Richard C. Dubes. *Algorithms for Clustering Data*. Prentice-Hall, 1988.

[15] Tony Lindeberg. *Scale-Space Theory in Computer Vision*. Int. Series in Engineering and Computer Science, Robotics: Vision, Manipulation and Sensors. Kluwer Academic Publishers, Dordrecht, 1994.

[16] Stephane G. Mallat. *A Wavelet Tour of Signal Processing*. Academic Press, Inc., 2nd edition, 2001.

[17] Pabitra Mitra, C. A. Murthy, and Sankar K. Pal. Density-based multiscale data condensation. *IEEE Trans. on Pattern Analysis and Machine Intelligence*, 24(6):734–747, June 2002.

Local Patterns: Theory and Practice of Constraint-Based Relational Subgroup Discovery

Nada Lavrač[1,2], Filip Železný[3], and Sašo Džeroski[1]

[1] Jožef Stefan Institute, Jamova 39, 1000 Ljubljana, Slovenia
[2] Nova Gorica Polytechnic, Vipavska 13, 5000 Nova Gorica, Slovenia
[3] Czech Technical University, Prague, Czech Republic

Abstract. This paper investigates local patterns in the multi-relational constraint-based data mining framework. Given this framework, it contributes to the theory of local patterns by providing the definition of local patterns, and a set of objective and subjective measures for evaluating the quality of induced patterns. These notions are illustrated on a description task of subgroup discovery, taking a propositionalization approach to relational subgroup discovery (RSD), based on adapting rule learning and first-order feature construction, applicable in individual-centered domains. It focuses on the use of constraints in RSD, both in feature construction and rule learning. We apply the proposed RSD approach to the Mutagenesis benchmark known from relational learning and a real-life telecommunications dataset.

1 Introduction

Inductive databases [11] embody a database perspective on knowledge discovery, where knowledge discovery processes are considered as query processes. In addition to normal data, inductive databases contain patterns (either materialized or defined as views). Data mining operations looking for patterns are viewed as queries posed to the inductive database. In addition to patterns (which are of local nature), models (which are of global nature) can also be considered.

A general formulation of data mining [19] involves the specification of a language of patterns and a set of constraints that a pattern has to satisfy with respect to a given database. The constraints that a pattern has to satisfy can be divided in two parts: language constraints and evaluation constraints. The first only concern the pattern itself, the second concern the validity of the pattern with respect to a database.

1.1 Constraints in Inductive Databases

Inductive queries consist of constraints and the primitives of an inductive query language include language constraints (e.g., find association rules with item A in the head) and evaluation primitives. Evaluation primitives are functions that express the validity of a pattern on a given dataset. We can use these to form evaluation constraints (e.g., find all item sets with support above a threshold)

K. Morik et al. (Eds.): Local Pattern Detection, LNAI 3539, pp. 71–88, 2005.
© Springer-Verlag Berlin Heidelberg 2005

or optimization constraints (e.g., find the 10 association rules with highest confidence).

Constraints thus play a central role in data mining and constraint-based data mining is now a recognized research topic [4]. The use of constraints enables more efficient induction as well as focussing the search for patterns on patterns likely to be of interest to the end user. While many different types of patterns have been considered in data mining, constraints have been mostly considered in mining frequent itemsets and association rules, as well as some related tasks, such as mining frequent episodes, Datalog queries, molecular fragments, etc. Few approaches exist that use constraints for other types of patterns/models, such as size and accuracy constraints in decision trees [10] or in classification rule discovery.

1.2   Constraints in Relational Subgroup Discovery

In this paper, we consider the use of constraints in the context of relational subgroup discovery (RSD). We consider the task of subgroup discovery defined as follows: given a population of individuals and a specific property of those individuals that we are interested in, find population subgroups that are statistically 'most interesting', e.g., are as large as possible and have the most unusual statistical (distributional) characteristics with respect to the property of interest [32]. We restrict ourselves to class-labeled data in our approach, with the class attribute being the property of interest.

While the goal of standard rule learning is to generate models, one for each class, inducing class characteristics in terms of properties occurring in the descriptions of training examples, in contrast, subgroup discovery aims at discovering individual 'patterns' of interest. In this sense, subgroup discovery belongs to *descriptive induction* [23,34] which has recently gained much attention of researchers developing rule learning algorithms. These involve mining of association rules (e.g., the APRIORI association rule learning algorithm [1]), clausal discovery (e.g., the CLAUDIEN system [23,24]), subgroup discovery (e.g., the MIDOS [32,33], EXPLORA [12], SD [9] and CN2-SD [17] subgroup discovery systems) and other approaches to non-classificatory induction aimed at finding interesting patterns in data.

Our approach to constraint-based RSD first performs feature generation, then applies a propositional approach to subgroup discovery (the RSD implementation in the Yap Prolog with a user's manual and sample problems can be obtained from http://labe.felk.cvut.cz/~zelezny/rsd). The combination of the above mentioned strategies controlled by constraints represents an original approach to relational subgroup discovery, although previous work exists incorporating some of the techniques mainly in classification rule discovery; e.g., rule induction with constraints in relational domains including propositionalization [2,3], or using rule sets to maximize ROC performance [7].

1.3   Outline of the Paper

This paper investigates local patterns in the multi-relational constraint-based data mining framework. Given this framework, it contributes to the theory of local patterns by providing the definition of local patterns and proposing a set of objective and subjective measures for evaluating the quality of induced patterns (Section 2). These notions are applied to a description task of subgroup discovery, for which a practical relational subgroup discovery algorithm RSD has been developed (Section 3). Section 4 discusses the use of constraints in RSD, followed by the experimental evaluation of the proposed approach to subgroup discovery in Section 5.

2   Theory of Local Patterns

This section contributes to the theory of local patterns by providing the definition of local patterns, and proposing a set of objective and subjective measures for evaluating the quality of induced patterns.

2.1   Pattern Discovery as Rule Learning

As in classification rule learning, we consider patterns of the form of a (backwards) implication:

$$Class \leftarrow Cond$$

Having limited the form of patterns to the above rule form, we limit the scope of investigation to patterns with a certain property of interest which is the goal of investigation (the target class, $Class$) that appears in the rule consequent. In the selected formalism the rule antecedent ($Cond$) is a conjunction of features (attribute-value pairs) selected from the features describing the training instances.

In the given scope, pattern discovery is a task at the intersection of predictive and descriptive induction. By inducing rules from labeled training instances (labeled positive if the property of interest holds, and negative otherwise), the process of subgroup discovery is targeted to uncovering properties of a selected target population of individuals with the given property of interest. In this sense, pattern discovery is a form of *supervised learning*. The fact that a pattern discovery task aims at characterizing population subgroups of a given target class suggests that standard classification rule learning could be used for solving the task. However, pattern discovery is a form of *descriptive induction* as the task is to uncover individual rules or patterns of interest, which must be represented in explicit symbolic form and which must be relatively simple in order to be recognized as actionable by potential users.

Each pattern can be extended with the information about the rule *quality*. Unlike in association rule learning, where rules are equipped with the *support*

and *confidence* of a rule, in this paper a standard rule pattern has the following
form:

$$Class \leftarrow Cond \; [TPr, FPr] \tag{1}$$

where $Class$ is the target property of interest, $Cond$ is a conjunction of features
(attribute-values), TPr is the *true positive rate* or the *sensitivity*, computed as
$p(Cond|Class) = \frac{n(Class.Cond)}{Pos}$, and FPr is the *false alarm* or *false positive rate*,
computed as $p(Cond|\overline{Class}) = \frac{n(\overline{Class}.Cond)}{Neg}$. In these formulas $n(Class.Cond)$
is the number of true positives TP (the number of covered instances belonging to
$Class$), $n(\overline{Class}.Cond)$ the number of false positives FP (the number of covered
instances not belonging to $Class$), Pos is the number of positives (instances of
the target class), Neg the number of negatives, and $N = Pos + Neg$ is the size
of the entire population.

2.2 Pattern Evaluation Measures

One can distinguish between *objective* and *subjective* quality measures (measures
of interestingness) [26]. Both the objective and subjective measures need to be
considered in order to solve pattern discovery tasks. Which of the quality criteria
are most appropriate depends on the application. Obviously, for automated rule
induction it is only the objective quality criteria that apply. However, for evaluating the quality of induced patterns and their usefulness for decision support,
the subjective criteria are more important, but also harder to evaluate.

As shown in Section 2.1, each rule can be extended with the information
about the rule quality. While the basic information of rule quality is usually
attached to the induced rule itself, as output of the learning algorithm, other
quality measures are usually computed for a ruleset, in order to evaluate the
output of the induction process as a whole, enabling the comparison of the
performance of different algorithms.

Below is a list of *subjective* measures of interestingness:

- *Usefulness.* Usefulness is an aspect of rule interestingness which relates a
  finding to the goals of the user [12].
- *Operationability.* In this paper we have introduced the notion of operationability, which is one aspect of usefulness.
- *Actionability.* "A rule is interesting if the user can do something with it to his
  or her advantage" [25,26]. Actionability is a special case of operationability.
- *Unexpectedness.* A rule is interesting if it is surprizing to the user [26].
- *Novelty.* A finding is interesting if it deviates from prior knowledge of the
  user [12].
- Redundancy. Redundancy amounts to the similarity of a finding with respect
  to other findings; it measures to what degree a finding follows from another
  one [12], or to what degree multiple findings support the same claims.

When discussing the *objective* quality measures - in line with the distinction between *predictive induction* and *descriptive induction* - one can distinguish

between the *predictive* and *descriptive* quality measures. A typical predictive quality measure, measuring the quality of a ruleset, is *predictive accuracy* of a ruleset, defined as the percentage of correctly predicted instances.[4]

In contrast with predictive quality measures, descriptive quality measures evaluate each individual subgroup and are thus appropriate for evaluating the success of pattern discovery. The following measures turn out to be most appropriate for measuring the quality of individual rules: rule size, coverage, support, accuracy (in different contexts also called precision or confidence), significance and unusualness. Although the evaluation of each individual rule is of ultimate importance, their variants that compute the average over the induced set of subgroup descriptions enable the comparisons of subgroup discovery algorithms (see [17] for the exact definition of these measures).

To explain rule significance and unusualness, which are the most important pattern discovery measures, some of the other measures for evaluating the quality of rules of the form $Class \leftarrow Cond$ need to be explained first. *Coverage* $p(Cond)$ is a measure of *generality*, computed as the relative frequency of all the examples covered by the rule: $\frac{n(Cond)}{N}$. *Support* $p(Class.Cond)$ is computed as the relative frequency of correctly classified covered examples: $\frac{n(Class.Cond)}{N}$. Rule *accuracy* $p(Class|Cond)$ (called *precision* in information retrieval and *confidence* in association rule learning) is the fraction of predicted positives that are true positives. Next, we define *accuracy gain* as the difference between rule accuracy $p(Class|Cond)$ and default accuracy $p(Class)$ achieved by the trivial rule $Class \leftarrow true$.

– *Significance* of a rule is computed in terms of the likelihood ratio of a rule, normalized with the likelihood ratio of the significance threshold (99%). Significance (or *evidence*, in the terminology of [12]) indicates how significant is a finding if measured by this statistical criterion. In the CN2 algorithm [5], significance $Sig(Class \leftarrow Cond)$ is measured in terms of the likelihood ratio statistic[5] of a rule as follows:

$$2 \sum_i n(Class_i.Cond) . \log \frac{n(Class_i.Cond)}{n(Class_i) \cdot p(Cond)} \tag{2}$$

where for each class $Class_i$, $n(Class_i.Cond)$ denotes the number of instances of $Class_i$ in the set where the rule body holds true, $n(Class_i)$ is the number of $Class_i$ instances, and $p(Cond)$ (i.e., rule coverage computed as $\frac{n(Cond)}{N}$) plays the role of a normalizing factor. Note that although for each generated subgroup description one class is selected as the target class, the significance criterion measures the distributional unusualness unbiased to any particular class – as such, it measures the significance of rule condition only: $Sig(Class \leftarrow Cond) = Sig(Cond)$.

---

[4] For a binary classification problem, ruleset accuracy is computed as $\frac{TP+TN}{N}$.
[5] In two-class problems this statistic is distributed approximately as χ^2 with one degree of freedom.

- *Unusualness* of a rule is computed by the *weighted relative accuracy* of a rule [15], defined as follows:

$$WRAcc(Class \leftarrow Cond) = p(Cond).[p(Class|Cond) - p(Class)]$$

Weighted relative accuracy can be understood as trading off rule *coverage* $p(Cond)$ and *accuracy gain* $p(Class|Cond) - p(Class)$.

As shown in [17], *WRAcc* is appropriate for measuring the unusualness of patterns, because it is proportional to the vertical distance from the diagonal in the ROC space (for ROC analysis, see [22]). As such, *WRAcc* also reflects rule significance - the larger *WRAcc* is, the more significant the rule is, and vice versa. As both *WRAcc* and rule significance measure the distributional unusualness of a pattern, they are the most important quality measures for pattern discovery, if the goal of pattern mining is— as is the case in this paper—finding of interesting population subgroups which are sufficiently large and distributionally unusual. However, while significance only measures distributional unusualness, computed in terms of correctly classified covered examples of all classes, *WRAcc* takes explicitly the rule coverage into the account, therefore we consider *unusualness* to be the most appropriate measure for pattern quality evaluation.

It can be shown that for a given pattern, its *WRAcc* value is proportional to the value of the Area Under the ROC Curve (*AUC*). Consequently, as optimizing *WRAcc* means also optimizing *AUC*, *WRAcc* proves to be of use not only as a heuristic appropriate for pattern discovery in descriptive induction, but also for predictive induction. This claim is supported by the results achieved in [29,16] in the comparisons of variants of CN2 and CN2-SD in which *WRAcc* was used instead of the rule accuracy heuristic.

3   Background: Relational Subgroup Discovery

Our approach adapts classification rule learning to relational subgroup discovery, described in [16], achieved by (a) propositionalization through first-order feature construction, (b) incorporation of example weights into the covering algorithm, (c) incorporation of example weights into the weighted relative accuracy search heuristic, (d) probabilistic classification based on the class distribution of covered examples by individual rules, and (e) area under the ROC curve rule set evaluation. The main advantage of the proposed approach is that each induced rule with a high weighted relative accuracy represents a 'chunk' of knowledge about the problem, due to the appropriate tradeoff between accuracy and coverage, achieved through the use of the weighted relative accuracy heuristic.

The input to the RSD algorithm consists of *a relational database* containing one main table (relation), where each row corresponds to a unique *individual* and one attribute of the main table is specified as the *class* attribute - this table defines the *training examples*, and other tables (relations) defining the *background knowledge*. In addition, *a mode-language definition* is given, which is used to construct first-order features.

The output of RSD is a set of subgroups whose class distributions differ substantially from the class distribution in the complete data set. The subgroups are defined by conjunctions of (automatically generated/defined) first-order features. The RSD algorithm proceeds in two stages: first-order feature construction and rule-based subgroup discovery.

RSD First-Order Feature Construction In our approach to first-order feature construction, based on [8,13,18], local variables referring to parts of individuals are introduced by so-called *structural predicates*. In a given language bias for first-order feature construction, a first-order feature is composed of one or more structural predicates introducing a new variable, and of *utility predicates* as in LINUS [14] (called *properties* in [8]) that 'consume' all new variables by assigning properties of individuals or their parts, represented by variables introduced so far. Utility predicates do not introduce new variables. (Examples of both types of predicates will be given below.)

The design of an algorithm for constructing first-order features can be split into two relatively independent problems:

Step 1: Identify features. This step results in identifying all first-order literal conjunctions that form a feature in the sense explained above, and at the same time comply to user-defined constraints (mode-language). Such features do not contain any constants and the task can be completed independently of the input data.

Step 2: Employ constants. This step results in extending the feature set by variable instantiations. Certain features are copied several times with some variables substituted to constants 'carefully' chosen from the input data. During this process, some irrelevant features are detected and removed, based on several constraints.

Both steps can be viewed as an exploitation of the combination of pre-set and user-defined sets of constraints of both syntactic (language-related) and semantic (data-oriented) character. From this viewpoint, they will be explained in detail in the devoted Section 4.

RSD Rule Induction Algorithm The core of RSD is a subgroup discovery algorithm which can accept data propositionalized by the feature constructor described above. The algorithm inherits some basic principles of the CN2 rule learner [5], which are adapted in several substantial ways to meet the needs of subgroup discovery. The principal improvements, making it appropriate for subgroup discovery, involve the implementation of the weighted covering algorithm, incorporation of example weights into the weighted relative accuracy heuristic, probabilistic classification, and the area under the ROC curve rule set evaluation [16].

4   Using Constraints in RSD

The curse of combinatorial dimensionality is present in the principles underlying both procedural phases of RSD:

- We apply language constraints to define the language of possible subgroup descriptions. These are applied both in feature generation and rule induction.
- We apply evaluation constraints during rule induction to select the (most) interesting rules/subgroups.

Consequently, RSD makes heavy use of both syntactic and semantic constraints exploited by search-space pruning mechanisms. On one hand, some of the constraints (such as *feature undecomposability*) are deliberately enforced by the system and pruning based on these constraints is guaranteed not to cause the omission of any solution. On the other hand, additional contraints (e.g. maximum *variable depth*) may be tuned by the user. These are designed with the intention to most naturally reflect possible user's heuristic expectations or minimum requirements on quantitative evaluations of search results.

4.1   Constraints in Feature Construction

Motivated by language-bias declarations used in ILP systems, RSD accepts language declarations very similar to those used by the systems Aleph [27] and Progol [21], including variable types, modes, setting a *recall* parameter etc, used to syntactically constrain the set of possible features. The use of the language bias declarations are best explained on a simple example. For this purpose we use the well-known East-West trains domain [20].

Structural predicates. By the mode declaration :-modeb(1,hasCar(+train, -car)) the user tells the system that the binary background relation hasCar may be employed in the body of constructed features, so as to provide the identification of some car of a specified train. The number 1 ("recall") determines that a feature can address at most one car of a given train. Input variables are labeled by the + sign, and output variables by the - sign.
Property predicates. Defined as above, but have no output variables.
Head predicate. Its declaration always contains exactly one variable of the input mode (e.g., :-modeh(1, train(+train)). The declaration serves merely to identify the key of the main individual.[6]

RSD produces all features satisfying the mode and setting declarations. The features produced by RSD have to satisfy an important constraint: a feature may not be decomposable into a conjunction of two features.

For example, the feature set based on the modes
```
:-modeh(1, train(+train)).
:-modeb(2, hasCar(+train, -car)).
```

---

[6] The head declaration thus may seem overly complicated but contributes to compatability with declarations used with the widely used ILP systems mentioned earlier.

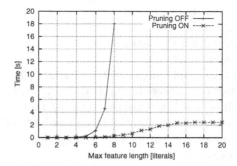

Fig. 1. The effect of pruning in the syntactic feature construction on efficiency in the East-West Trains domain. The diagram shows the amount of time needed to produce the exhaustive set of features for a given maximum feature length when pruning is off or on.

```
:-modeb(1, long(+car)).
:-modeb(1, notSame(+car, +car)).
```
will contain a feature
```
f(A):- hasCar(A,B),hasCar(A,C),long(C),long(B),notSame(B,C).
```
but it will not contain a feature with the body
```
hasCar(A,B),hasCar(A,C),long(B),long(C)
```
as such an expression would clearly be decomposable into two separate features. We do not construct such decomposable expressions, as these are redundant for the purpose of the subsequent search for rules with conjunctive antecedents.

The language constraint of undecomposability plays a major role: it enables pruning the search for possible features without losing any solutions. As an example, Figure 1 illustrates the speedup gained by the pruning on the East-West Trains domain (evaluation on real-life data will be shown in the experimental section).

In addition, other language constraints can be specified. These are: the maximum length of a feature (number of contained literals), maximum *variable depth* [21] and maximum number of occurrences of a given predicate symbol. If constraints are not specified by the user, the first two acquire a default value while the last is unlimited.

Unlike Aleph and Progol declarations, RSD does not use the # sign to denote a constant-value argument. In the mentioned systems, constants are provided by a single saturated example, while RSD extracts constants from all the input data (examples). The user can instead utilize the special reserved property predicate `instantiate/1`, which does not occur in the background knowledge, to specify a variable that should be substituted with a constant during feature construction. For example, from the modes
```
:-modeh(1, train(+train)).
:-modeb(1, hasCar(+train, -car)).
```

```
:-modeb(1, hasLoad(+car, -load)).
:-modeb(1, hasShape(+load, -shape)).
:-modeb(*, instantiate(+shape)).
```
exactly one feature is generated:

```
f1(A) :- hasCar(A,B), hasLoad(B,C), hasShape(C,D), instantiate(D).
```
In the second step, after consulting the input data, `f1` will be substituted by a set of features, in each of which the `instantiate/1` literal is removed and the D variable is substituted with a constant making the body of `f1` provable in the data. Provided they contain a train with a rectangle load, the following feature will appear among those created out of `f1`:

```
f11(A) :- hasCar(A,B), hasLoad(B,C), hasShape(C,rectangle).
```
A similar principle applies for features with multiple occurences of the `instantiate/1` literal. Arguments of this literal within the feature form a set of variables ϑ; only those (complete) instantiations of ϑ making the feature's body provable on the input database will be considered.

However, not all such features will appear in the resulting set. For the sake of efficiency, we do not perform feature filtering by a separate postprocessing procedure, but rather discard certain features already during the feature construction process described above. The following constraints are used: (a) no feature should have the same value for all examples and (b) no two features should have the same values for all examples. For the latter case, only the syntactically shortest feature is chosen to represent the class of semantically equivalent features. In addition, a minimum number of examples for which a feature has to be true can be prescribed. This constraint is similar to the minimum support constraints in mining frequent item sets.

4.2 Constraints in Subgroup Discovery

In the subgroup discovery phase, a language constraint employed is the prescription of a maximal number of conditions/features in the description of a subgroup.

Several evaluation functions are considered. These include accuracy, weighted relative accuracy (*WRAcc*), significance, and area under the ROC curve. Accuracy and *WRAcc* are used in optimization constraints, i.e., RSD looks for rules with high accuracy/*WRAcc*. In fact, they are used as heuristic functions in RSD. Significance is used in evaluation constraints, i.e., one can prescribe a significance threshold that rules have to satisfy (expressed as significance at e.g., 99% level). *WRAcc* may be used in a similar fashion.

For lack of space we do not provide here a tabular summary of all employed constraints and the ways of their setting, this can be however found in the RSD user's manual available from the above mentioned RSD download page.

5   Experiments

We have experimented with the well-known relational learning benchmark concerning the Mutagenicity of chemicals and we have also applied RSD to the analysis of a real-life telecommunications dataset. The Mutagenesis data have been described in detail in many sources, see e.g. [28]. The Telecommunication application has been described by Železný et al. in [31,30]; next we give a brief overview of the data.

5.1 Telecommunications

The data represent incoming calls (1995 items thereof) to an enterprize. Each such call is answered by a human operator and in the usual case further transferred to an attendant distinguished by his/her line number. Further re-transfers may also occur. Each sequence of such transfers is tracked by a computerized exchange and related data are stored in a logging file. By a suitable transformation thereof, one may obtain a relation `incoming/5`, represented by ground facts of the form `incoming(`*date, time, caller, operator, result*`)`. The argument *result* either takes a constant value or is a recursively defined function, so that *result* $\in \{$`talk`, `unavailable`, `transfer([`$ln_1, ln_2, ..., ln_n$`]`, *result*`)}`, where $ln_1...ln_{n-1}$ denote line numbers to which unsuccessful attempts to transfer have been made, and ln_n the result of the last transfer attempt.

For example, the ground fact
```
incoming(date(10,18), time(13,37,29), [0,6,4,8,2,5,6,8,4,9], 32,
   transfer([16,12],transfer([26],talk))).
```
describes a call from the number 0648256849 at 13:37:29 on 10/18 received by the operator on line 32. The operator first tried to transfer the caller to line 16 without success, and then transferred him/her successfully to line 12. The person on line 12 further redirected the caller to line 26. After a talk with line 26, the call was terminated.

We divide all instances of incoming transferred calls into classes determined by the line to which the operator tried to transfer the caller first. We thus obtain 25 classes. Attributes of examples (the main table records) then consist of the first four arguments of `incoming/5` and the class attribute. Finding subgroups interesting with respect to this class attribute may contribute to purposes of decision support of the operator. Further, if the subgroup set has sufficient predictive power, it may partially or completely substitute the operator.

Let us now comment on two of the available background relations. The predicate `prefix(Number,Prefix)` is true whenever the second (output) argument is the prefix (of any length) of the first (input) argument. For instance, regarding the example given above, `prefix([0,6,4,8,2,5,6,8,4,9],[0,6,4])` is true.

This background predicate proved useful in previously published results, since it is able to bind callers from the same area, city, company, office etc. The predicate gives multiple possible outputs for a given input. When used as part of a feature definition, it will be the job of the feature constructor to decide which prefixes should be used (possibly in conjunction with other literals) to generate features with acceptable coverage measures. Out of the prefixes kept, the rule inducer chooses those that help identify interesting subgroups.

Another background predicate `prev_attempt/6` reflects the fact that a line desired by the caller may often be determined by looking at the caller's recent attempts to reach a person, i.e., by inspecting past records (w.r.t. the time-label of the current example) in the `incoming/4` relation. This problem setting is thus not far from what is known as *multi-instance learning* [6], where relevant attribute values describing an instance extend in multiple rows of a single table.

For example, the goal

```
prev_attempt(date(10,18),time(13,37,29),
[0,6,4,8,2,5,6,8,4,9], Line, When, Result).
```

will succeed with the result

```
Line=10, When=today, Result = unavailable,
```

provided the caller 0648256849 failed to reach line 10 on 10/18 before 13:37:29. Again, the `prev_attempt/6` may obviously yield multiple outputs for a given instantiations of the input arguments.

5.2   Expert Analysis of Induced Subgroups: Evaluating Novelty

We present the descriptions of some of the discovered subgroup in Telecommunication, with comments from the domain expert on the descriptions in Table 1 and the distributional characteristics of the subgroups.

Expert analysis of the induced rules shows that some of them identify novel and interesting information. Especially revealing are the comments related to the changes of class frequency associated with the rules. In the overall distribution, calls to line 21 are most common. The expert comments that this reflects his expectations, as the person at line 21 is a marketer, and people interested in products call this line most frequently. In subgroup `Tele1`, there is (a) an increase in line 21 frequency: clients not receiving an ordered package often wait until Friday and then complain with line 21; and (b) a decrease in line 13 frequency: the person at line 13 mostly collaborates with dealers who have less business on Fridays. For subgroup `Tele4` there is a) an increase in line 28 frequency: repeated attempts to reach line 28, and (b) an increase in line 21 frequency: the person at line 28 works as technical support for products sold by person on line 21.

The use of the undecomposability constraint and the pruning enabled thereby greatly reduces the time necessary to generate the features. This reduction increases with the maximum feature length, as illustrated in Figure 2.

5.3   Effects of Constraints on Feature Generation

The use of the other feature constraints, i.e., the minimum coverage, unique coverage and incomplete coverage (the latter two are referred to as filtering) reduces the number of features generated, as shown in Figure 3. In Mutagenesis, the maximum feature length was set to 5 and the minimum feature coverage to 20 instances, obtaining 42 different features. In the Telecom domain, we set the maximum feature length to 8. In this case, using a minimum coverage of 20 instances yields 138 features.

Table 1. Subgroup descriptions in the form $H \leftarrow B$ $[TP, FP]$, definitions of used features, and subgroup interpretation including expert's comments.

Tele1: `line21(A) ← f40(A)` [56,268]
`f40(A):-call_date(A,B),dow(B,fri).`
Calls received on Fridays.
Expert's evaluation: Not a novel information.

Tele2: `line11(A) ← f132(A)` [32,0]
`f132(A):-ext_number(A,B), prefix(B,[8,5,1,3,1,1,1,1]).`
Calls received from number 85131111.
Expert's explanation: The caller is the secretary's husband. She does not have a direct-access line, thus this call is transferred by an operator.
Expert's evaluation: Novel information.
Remark. Although the last literal formally identifies a *prefix* of the calling number, it is in fact the complete number of the caller.

Tele3: `line21(A) ← f54(A)` [81,254]
`f54(A):-ext_number(A,B),prefix(B,[0,4]).`
Calls received from a number that starts with 04.
Expert's explanation: Prefix 04 is too general (code covers a large area) to find an explanation.
Expert's evaluation: Novel information. Uncertain.

Tele4: `line28(A) ← f7(A)` [22,11]
`f7(A):-call_date(A,B),call_time(A,C),`
    `ext_number(A,D), prev_attempt(B,C,D,[2,8], last_hour, unavailable).`
Calls received from a caller who has in the last hour attempted to directly (not through an operator) reach line 28, which was unavailable.
Expert's explanation: It is plausible that people try line 28 as the second attempt when line 21 is unavailable. Subgroup probably mostly covers people with technical difficulties with a product sold by person on line 21.
Expert's evaluation: Novel information.

5.4 Results of Subgroup Discovery

An example feature in the Mutagenesis domain is `f12(A):-atm(A,B),atm_chr(B,C),lteq_c(C,0.142)` expressing that a drug contains an atom with charge less or equal to 0.142, or `f31(A):-benzene(A,B),benzene(A,C), connected(C,B)`, expressing the presence of two connected benzene rings in the chemical. In telecommunications, an example feature is `f99(A):-ext_number(A,B),prefix(B,[0,4,0,7])`, meaning that the caller's number starts with 0407. Another feature is `f115(A):- call_date(A,B),call_time(A,C),ext_number(A,D), prev_attempt(B,C,D,[3,1], today,unavailable)`, meaning that the caller (of the current call) has today tried to reach line 31, which was unavailable.

With these features, we use the RSD rule induction algorithm with altered covering strategy and heuristic function to produce sets of subgroup-describing rules.

The characteristics of the discovered rules are shown in Table 2. Algo refers to the combination of search heuristic (A-accuracy, W-WRacc (weighted relative accuracy)) and covering algorithm (C-covering, W-WeCov (weighted covering

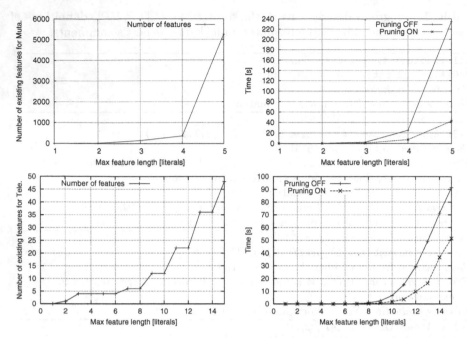

Fig. 2. The number of existing features (left) and the effect of undecomposability enabled pruning in the syntactic feature construction on efficiency (right) in the Mutagenesis (top) and Telecommunication (bottom) domain.

with $\gamma = 1$)). S = significance, C = coverage, A = area under ROC curve. R : F = average number of rules/class : average number of features per rule. R′ : F′ = same as above, only rules on covex hull are considered. The rule generation for a given class is terminated if the search space has been completely explored or 10 subgroup rules have been generated for that class in Telecommunication (5 in Mutagenesis). Reported results are averages and standard deviations from a 10-fold stratified cross-validation procedure.

The most significant observation about the results in Table 2 is that the *WRAcc* heuristic very significantly improves the performance with respect to the other accuracy heuristics, in terms of all three quality aspects.

Overall, the combination of *WRAcc* with the strategy of example weighting yields the best performance. This agrees with the findings in [17], where a more extensive empirical evaluation was conducted on a collection of (non-relational) subgroup-discovery problems, comparing the CN2 algorithm with CN2 incorporating the *WRAcc* heuristic, and further the CN2-SD system (which incorporates the *WRAcc* heuristic and the example weights). These three algorithms roughly correspond to the methods we denote above (in Table 2) as AC, WC, and WW, respectively. The combination of the accuracy heuristic with example weighting (AW) seems not to perform well in the domains considered.

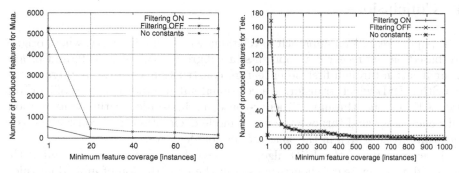

Fig. 3. The effect of using the constraint of unique and incomplete coverage ("Filtering ON" line) vs. ignoring this constraint ("Filtering OFF" line) and the user-adjustable minimum-coverage constrain (left-to-right decay) for Mutagenesis (left) and Telecommunication (right). The "No constants" curve (independent of the horizontal axis) corresponds to the number of features before instantiations to constants (before Step 2 of feature construction), this number is high in Mutagenesis due to many features inprovable with any instantiation to constants.

Table 2. Characteristics of subgroup-describing rules obtained by the RSD rule induction algorithm in the Mutagenesis and Telecom domains.

Mutagenesis

| Algo | Performance | | | Complexity | |
	S	C	A	R : F	R' : F'
AC	1.99	11.33%	0.69	10.00 : 2.16	10.00 : 2.16
	(0.92)	(3.74)	(0.07)	(0.00 : 0.07)	(0.00 : 0.07)
AW	1.33	7.62%	0.58	10.00 : 2.50	10.00 : 2.50
	(1.05)	(4.88)	(0.06)	(0.00 : 0.11)	(0.00 : 0.11)
WC	4.22	35.81%	0.86	3.70 : 1.73	2.30 : 1.62
	(1.22)	(6.44)	(0.06)	(0.82 : 0.33)	(0.48 : 0.22)
WW	7.48	40.58%	0.90	10.00 : 2.63	6.50 : 2.43
	(1.28)	(4.74)	(0.04)	(0.00 : 0.07)	(0.97 : 0.11)

Telecommunication

| Algo | Performance | | | Complexity | |
	S	C	A	R : F	R' : F'
AC	2.90	0.37%	0.55	7.36 : 2.39	6.88 : 2.47
	(0.38)	(0.05)	(0.02)	(0.12 : 0.04)	(0.19 : 0.04)
AW	2.25	0.25%	0.55	9.96 : 2.56	9.60 : 2.61
	(0.52)	(0.04)	(0.02)	(0.07 : 0.03)	(0.07 : 0.03)
WC	11.29	4.98%	0.67	6.12 : 2.17	5.20 : 2.28
	(1.71)	(0.54)	(0.02)	(0.16 : 0.04)	(0.16 : 0.04)
WW	11.99	4.02%	0.70	9.64 : 2.06	6.68 : 2.29
	(1.05)	(0.41)	(0.01)	(0.12 : 0.01)	(0.20 : 0.03)

6   Conclusions

This paper presents an approach to relational subgroup discovery, whose origins are based on the recent developments in subgroup discovery [33,9] and propositionalization through first-order feature construction [8,13,18]. It presents the algorithm RSD which transforms a relational subgroup discovery problem to a propositional one, through efficiency-conscious first-order feature construction. Efficiency is boosted through the use of mode declarations and constraints used for pruning the search in the space of possible features.

Four variants of the RSD algorithm have been tested, by combining the standard accuracy search heuristic used in the construction of individual rules, with the standard covering algorithm used in the construction of a set of rules. The *WRAcc* heuristic combined with the weighted covering algorithm is the preferred combination (due to an appropriate tradeoff between rule significance, coverage and complexity).

We have successfully applied the RSD algorithm in the Mutagenesis benchmark and the Telecom domain, a real-life dataset from a telecommunications company. These results have been evaluated as meaningful by the domain expert. Both the description of subgroups and their distributional characteristics make sense in many cases.

The idea of incrementally extending the feature set in dependence on the quality of the discovered subgroups, seems very much worth investigating in further work.

Acknowledgments

We are grateful to Peter Flach for the collaboration in the development of the CN2-SD algorithm and in the genesis phase of the RSD algorithm. We are grateful also to Jiří Zídek, Atlantis Telecom s.r.o. for the comments on the Telecom subgroups discovered. Nada Lavrač and Sašo Džeroski acknowledge the support of the Slovenian Ministry of Education, Science and Sport and the cInQ (Consortium on discovering knowledge with Inductive Queries) project, funded by the European Commission. Filip Železný was supported by the DARPA EELD grant F30602-01-2-0571 and the Czech Ministry of education grant MSM 21230013.

References

1. R. Agrawal, H. Mannila, R. Srikant, H. Toivonen and A.I. Verkamo. Fast discovery of association rules. In U.M. Fayyad, G. Piatetsky-Shapiro, P. Smyth and R. Uthurusamy, editors, *Advances in Knowledge Discovery and Data Mining*, 307–328, MIT Press, 1996.
2. J.M. Aronis and F.J. Provost. Efficiently constructing relational features from background knowledge for inductive machine learning. In U.M. Fayyad and R. Uthurusamy, editors, *AAAI Workshop on Knowledge Discovery in Databases*, 347–358, AAAI Press, 1994.

3. J.M. Aronis, F.J. Provost and B.G. Buchanan. Eploiting background knowledge in automated discovery. In E. Simoudis, J. Han and U. Fayyad, editors, *Proceedings of the Second International Conference on Knowledge Discovery and Data Mining*, 355–358, AAI Press, 1996.

4. R. Bayardo, editor. Constraints in data mining. Special Issue of *SIGKDD Explorations*, 4(1), 2002.

5. P. Clark and T. Niblett. The CN2 induction algorithm. *Machine Learning*, 3: 261–283, 1989.

6. L. De Raedt. Attribute value learning versus inductive logic programming: The missing links (extended abstract). In D. Page, editor, *Proceedings of the 8th International Conference on Inductive Logic Programming*, 1–8. Springer, 1998.

7. T. Fawcett. Using rule sets to maximize ROC performance. In *Proceedings of the 2001 IEEE International Conference on Data Mining*, 131–138, 2001.

8. P. Flach and N. Lachiche. 1BC: A first-order Bayesian classifier. In S. Džeroski and P. Flach, editors, *Proceedings of the 9th International Workshop on Inductive Logic Programming*, 92–103. Springer, 1999.

9. D. Gamberger and N. Lavrač. Expert guided subgroup discovery: Methodology and application. *Journal of Artificial Intelligence Research*, 17: 501–527, 2002.

10. M. Garofalakis and R. Rastogi. Scalable data mining with model constraints. *SIGKDD Explorations*, 2(2):39–48, 2000.

11. T. Imielinski and H. Mannila. A database perspective on knowledge discovery. *Communications of the ACM*, 39(11): 58–64, 1996.

12. W. Klösgen. Explora: A multipattern and multistrategy discovery assistant. In U.M. Fayyad, G. Piatetsky-Shapiro, P. Smyth and R. Uthurusamy, editors, *Advances in Knowledge Discovery and Data Mining*, 249–271, MIT Press, 1996.

13. S. Kramer, N. Lavrač and P. Flach. Propositionalization approaches to relational data mining. In S. Džeroski and N. Lavrač, editors, *Relational Data Mining*, 262–291. Springer, 2001.

14. N. Lavrač and S. Džeroski. *Inductive Logic Programming: Techniques and Applications*. Ellis Horwood, 1994.

15. N. Lavrač, P. Flach and B. Zupan. Rule evaluation measures: A unifying view. In: S. Džeroski and P. Flach, editors, *Proceedings of the 9th International Workshop on Inductive Logic Programming*, 174–185, Springer, 1999.

16. N. Lavrač, P. Flach, B. Kavšek and L. Todorovski. Adapting classification rule induction to subgroup discovery. In V. Kumar et al., editors, *Proceedings of the IEEE International Conference on Data Mining*, 266–273. IEEE Computer Society, December 2002.

17. N. Lavrač, B. Kavšek, P. Flach and L. Todorovski. Subgroup discovery with CN2-SD. *Journal of Machine Learning Research*, 5: 153–188, 2004.

18. N. Lavrač and P. Flach. An extended transformation approach to inductive logic programming. *ACM Transactions on Computational Logic*, 2(4): 458–494, 2001.

19. H. Mannila and H. Toivonen. Levelwise search and borders of theories in knowledge discovery. *Data Mining and Knowledge Discovery*, 1(3): 241–258, 1997.

20. D. Michie, S. Muggleton, D. Page and A. Srinivasan. To the international computing community: A new East-West challenge. Technical report, Oxford University Computing laboratory, Oxford, UK, 1994.

21. S. Muggleton. Inverse entailment and Progol. *New Generation Computing*, Special issue on Inductive Logic Programming, 13(3–4): 245–286, 1995.

22. F.J. Provost and T. Fawcett. Robust classification systems for imprecise environments. In *Proceedings of the Fifteenth National Conference on Artificial Intelligence and Tenth Innovative Applications of Artificial Intelligence Conference*, 706–713, MIT Press, 1998.
23. L. De Raedt and L. Dehaspe. Clausal discovery. *Machine Learning*, 26: 99–146, 1997.
24. L. De Raedt, H. Blockeel, L. Dehaspe and W. Van Laer. Three companions for data mining in first order logic. In S. Džeroski and N. Lavrač, editors, *Relational Data Mining*, 105–139. Springer, 2001.
25. G. Piatetsky-Shapiro and C. Matheus. The interestingness of deviation. In: *Proceedings of the AAAI-94 Workshop on Knowledge Discovery in Databases*, 25–36, 1994.
26. A. Silberschatz and A. Tuzhilin. On subjective measures of interestingness in knowledge discovery. In: *Knowledge Discovery and Data Mining*, 275–281, 1995.
27. A. Srinivasan and R.D. King. Feature construction with inductive logic programming: A study of quantitative predictions of biological activity aided by structural attributes. In S. Muggleton, editor, *Proceedings of the 6th International Workshop on Inductive Logic Programming*, 89–104. Springer, 1996.
28. A. Srinivasan, S. Muggleton, M.J.E. Sternberg and R.D. King. Theories for mutagenicity: A study in first-order and feature-based induction. *Artificial Intelligence*, 85(1–2): 277-299, 1996.
29. L. Todorovski, P. Flach and N. Lavrač. Predictive performance of weighted relative accuracy. In D.A. Zighed, J. Komorowski and J. Zytkow, editors, *Proceedings of the 4th European Conference on Principles of Data Mining and Knowledge Discovery*, 255–264. Springer, 2000.
30. F. Železný, P. Mikšovský, O. Štěpánková, and J. Zídek. ILP for automated telephony. In J. Cussens and A. Frisch, editors, *Proceedings of the Work-in-Progress Track of the 10th International Conference on Inductive Logic Programming*, 276–286, 2000.
31. F. Železný, J. Zídek, and O. Štěpánková. A learning system for decision support in telecommunications. In *Proceedings of the First International Conference on Computing in an Imperfect World, Belfast 4/2002*. Springer, 2002.
32. S. Wrobel. An algorithm for multi-relational discovery of subgroups. In J. Komorowski and J. Zytkow, editors, *Proceedings of the First European Symposion on Principles of Data Mining and Knowledge Discovery*, 78–87, Springer, 1997.
33. S. Wrobel. Inductive logic programming for knowledge discovery in databases. In S. Džeroski and N. Lavrač, editors, *Relational Data Mining*, 74–101. Springer,2001.
34. S. Wrobel and S. Džeroski. The ILP description learning problem: Towards a general model-level definition of data mining in ILP. In K. Morik and J. Herrmann, editors, *Proceedings Fachgruppentreffen Maschinelles Lernen (FGML-95)*, 44221 Dortmund, Univ. Dortmund, 1995.

Visualizing Very Large Graphs Using Clustering Neighborhoods

Dunja Mladenic and Marko Grobelnik

Jozef Stefan Institute, Jamova 39, 1000 Ljubljana, Slovenia
{Dunja.Mladenic, Marko.Grobelnik}@ijs.si
http://kt.ijs.si/Dunja/, http://kt.ijs.si/Marko/

Abstract. This paper presents a method for visualization of large graphs in a two-dimensional space, such as a collection of Web pages. The main contribution here is in the representation change to enable better handling of the data. The idea of the method consists from three major steps: (1) First, we transform a graph into a sparse matrix, where for each vertex in the graph there is one sparse vector in the matrix. Sparse vectors have non-zero components for the vertices that are close to the vertex represented by the vector. (2) Next, we perform hierarchical clustering (eg., hierarchical K-Means) on the set of sparse vectors, resulting in the hierarchy of clusters. (3) In the last step, we map hicrarchy of clusters into a two-dimensional space in the way that more similar clusters appear closely on the picture. The effect of the whole procedure is that we assign unique X and Y coordinates to each vertex, in a way those vertices or groups of vertices on several levels of hierarchy that are stronger connected in a graph are place closer in the picture. The method is particular useful for power distributed graphs. We show applications of the method on real-world examples of visualization of institution collaboration graph and cross-sell recommendation graph.

1 Introduction

When trying to find some regularity in the data, one can apply different methods depending on specific goals, data properties, resources etc. Visualization of the data is welcome in most cases and it sometimes can be of great help in getting top-level (zoomed-out) or local (zoomed in) insights in the data. Most of the methods developed in the field of Knowledge Discovery in Databases [1], [2], [6], [7], [13] are focused on finding global regularities that explain most of the instances. Recently, it was recognized that there is a lack of recognition of importance of local patterns and thus a lack of appropriate methods for finding local patterns [5]. There are several attempts to define local patters, for the purpose of clarification, let's just say that local patterns can potentially be found in the data as a kind of outliers or regularities on a local level that do not propagate to the global level (thus not capture by a global models of the data).

This paper addresses a problem of finding regularities including such as local patterns in the data represented as a graph. We propose a novel graph visualization ap-

K. Morik et al. (Eds.): Local Pattern Detection, LNAI 3539, pp. 89–97, 2005.

proach especially suitable for very large graphs. We see graph as a fundamental data structure and apply data-analytic techniques for dealing with graphs. Due to the visual nature of the results, it is difficult to impose a strict evaluation method on our approach. Thus we provide illustrative examples of our graph visualization results.

Related work deals mainly with relaxation methods minimizing some kind of ad-hoc aesthetic function which usually work fine for up-to several hundreds vertices, but to the best of our knowledge, there is no efficient generic solution for visualization of large graphs. Good source of related approaches is an annual "Graph Drawing Competition" [8],[9].

The paper is structured as follows. Section 2 provides problem and approach description. Section 3 gives example results on two large real-world graphs. Discussion is provided in Section 4.

2 Problem and Approach Description

The problem addressed in this paper is visualization of very large graphs, where we are given a graph with typically over 10,000 vertices. The goal is to draw the graph in a reasonable time and draw it nicely using some ad-hoc visual aesthetics. Our emphasis is on proposing an appropriate approach with the required time complexity, while evaluation of the aesthetics of the graphs is out of the scope of this paper.

The main idea of the proposed approach comes from the area of document collection visualization, where we have developed a couple of methods for visualization of hierarchically organized set of documents [3]. The important step that we propose here is to transform the given graph into a set of sparse vectors, where for each vertex in the graph we create one sparse vector (Section 2.1). On that set of sparse vectors we then perform hierarchical clustering, so that each leaf in the hierarchy includes just one instance. For hierarchical clustering we use bisecting k-means that is an efficient top-down approach to clustering (Section 2.2). The next important step that we have originally developed and tested on document collection visualization is to apply "tiling" technique on the hierarchy of clusters. The important property of "tiling" is that each instance in the hierarchy leaf gets X and Y coordinates uniquely assigned (Section 2.3). This combination of "tiling" and clustering performs the same function as multidimensional scaling used in traditional approaches. Finally, based on the assigned X and Y coordinates, we draw the graph in $2D$ space on the screen.

2.1 Transforming Graph into Sparse Vectors

In order to get sparse vectors from a graph, we represent a graph with N vertices as a NxN sparse matrix. The matrix is constructed so that the Xth row gives information about the Xth vertex and has nonzero components for the columns representing vertices from the neighborhood of vertex X. We have defined neighborhood of a vertex to contain all the vertices at the distance of up to d steps from the vertex. Consequently, the Xth row has non-zero component in the Xth column and all the other columns that represent the neighbors at step $1, 2, 3, \ldots d$. Intuitively, the Xth row numerically represents the neighborhood of the *Xth* vertex within the graph, with the values calculated using the following formula $1/2^d$, where d is the distance in the number of steps from the *Xth* vertex. Figure 1 illustrates the graph transformation on an example graph.

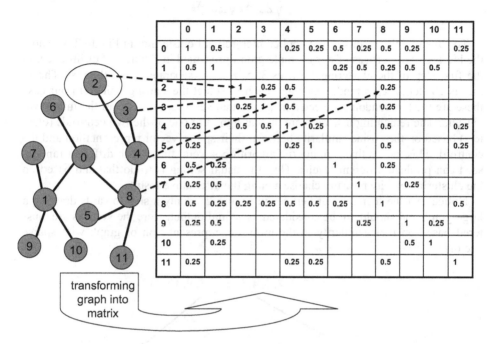

	0	1	2	3	4	5	6	7	8	9	10	11
0	1	0.5			0.25	0.25	0.5	0.25	0.5	0.25		0.25
1	0.5	1					0.25	0.5	0.25	0.5	0.5	
2			1	0.25	0.5				0.25			
3			0.25	1	0.5				0.25			
4	0.25		0.5	0.5	1	0.25			0.5			0.25
5	0.25				0.25	1			0.5			0.25
6	0.5	0.25					1		0.25			
7	0.25	0.5						1		0.25		
8	0.5	0.25	0.25	0.25	0.5	0.5	0.25		1			0.5
9	0.25	0.5						0.25		1	0.25	
10		0.25								0.5	1	
11	0.25				0.25	0.25			0.5			1

transforming
graph into
matrix

Fig. 1. Illustration of the graph transformation into a sparse matrix where the rows represent instances (vertices) and columns represent neighborhood with weights relative to the distance from the vertex in that row. Here we have set the maximal distance to d= 2. Notice that the diagonal elements have weight 1 (showing that the each vertex is in its own neighborhood). The dashed lines point out neighboring vertices and the corresponding weights for vertex labeled as 2. It has four non-zero elements in its sparse vector representation (1, 0.25, 0.5, 0.25) corresponding to four vertices (labeled in the graph as 2, 3, 4, 8)

2.2 Hierarchical Clustering

We perform hierarchical clustering on the set of sparse vectors representing graph vertices using top-down bisecting k-means computationally efficient, as applied for document clustering [12]. Starting with all the instances being in one cluster, we split the cluster into two clusters using 2-means clustering (see Figure 2 for illustration of the clustering results). Next the same procedure is repeated for each of the two newly obtained clusters and recursively further down the hierarchy. We perform divisive clustering until the size of the clusters is one (one instance per cluster). Each cluster on the transformed graph data represents an approximate clique in the graph; therefore we can view the clustering of vertices as generating hierarchy of approximate cliques in the original graph. We base a distance between the sparse vectors on cosine similarity, a measure commonly used in information retrieval and text-mining. Cosine similarity between the two vectors D_1, D_2 is defined as follows,

$$Sim(D_1, D_2) = \frac{\sum_i x_{1i} x_{2i}}{\sqrt{\sum_j x_j^2} \sqrt{\sum_k x_k^2}} . \tag{1}$$

To illustrate the results of the applied hierarchical clustering, in Figure 2 we show the hierarchy of vertices of our example graph presented in Figure 1. For instance, in the first step of the clustering, vertices 3, 5, 8 were separated from the rest. The k-means clustering uses randomized approach to select the starting points. In our case these are the two randomly selected instances that serve as centroids of the future two clusters. The remaining instances are then assigned to one of the two centroids (clusters) based on the cosine similarity between their sparse vector representation and the centroid. This means that running the algorithm several times with different random seed can produce different results. Thus the usual setting is to do that and based on the clustering quality measure choose among the alternative results.

Notice that the hierarchy of vertices in Figure 2 is only a step towards drawing a large graph and should not be considered as a final output. Here the vertices are clustered based on their similarity in the proposed representation of graph with sparse vectors.

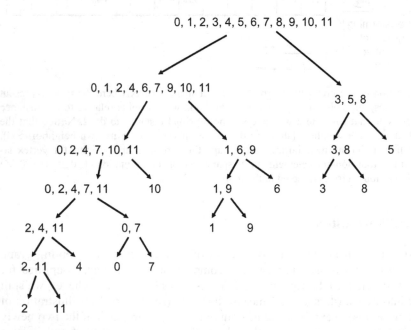

Fig. 2. The result of our hierarchical clustering applied on sparse vectors representing the graph vertices obtained on the example graph given Figure 1

2.3 Tiling Based Visualisation

The tilling technique is known for structured data from data visualization community, eg., [11]. We have previously adapted the technique for textual data to enable efficient visualization of document collection [3]. Here, we apply the same technique for visualization of graphs as follows.

Given is the hierarchy of instances and a fixed size two dimensional area (on the screen). The main idea is to split the rectangular viewing area into sub-areas according to the size of the clusters. In the case of visualizing a hierarchy of clusters, we generate an image on several levels, giving a top (bird) view on the whole hierarchy. The technique takes care that each cluster has uniquely determined position in the picture and in the extreme case, where the size of clusters is 1, each cluster (each individual instance) has determined a numerical X and Y coordinates. We use this effect of explicitly assigning coordinates to the instances to position vertices in a graph. Since our graph vertices are grouped using clustering based on their closeness in the original graph, we expect that also the instances (representing the original graph vertices) on the picture will stay close, if they are neighbor in the original graph.

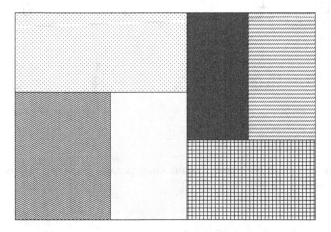

Fig. 3. Illustration of tilling based visualization approach. Each block includes set of instances with blocks having similar instances placed closer in the picture. In the first step the vertical division of the area is performed in the left part (in this illustration containing about 60% of the instances) and the right part (in this illustration containing about 40% of the instances). Then each of the two parts is further split horizontally and then vertically (then, assume that the stopping criterion is met, so we stop the process)

Starting with the root node of the hierarchy the whole rectangular drawing area partitioned to the both children proportionally to the number of instances assigned to them via hierarchical clustering (see Figure 3 for illustration of the area partitioning). This procedure is performed recursively down to the leaf nodes. In general, in tilling based visualization, we stop the splitting of the drawing area based on some stopping criterion, such as the minimum number of instances in the cluster. While in document visualization, we usually stop when the number of documents in the cluster is higher then one (maybe 10% of the document collection size), in the case of graph drawing the stopping criterion is set to having one instance in the cluster. Namely, only in this way we can at the end assign a distinct position to each of the graph vertices.

Figure 4 illustrates the result of our approach on drawing our example graph. Notice that for Figure 1, the same graph was manually drawn with nice placement of the vertices just for presenting it in the paper, but that information is not provided to the system. The system is given only usual, text representation of the graph, listing vertices and their neighbors, as shown in Figure 4. Notice that vertices that are in the same cluster (eg., 3,8) are not necessary placed close to each other.

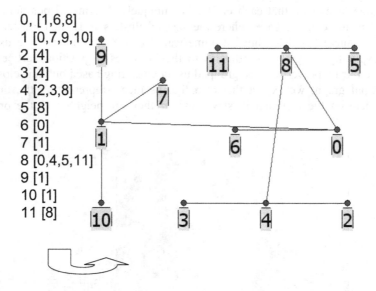

Fig. 4. The result of graph visualization of the same example graph that was used for illustration of the approach in Figure 1 and Figure 2

3   Results on Two Large Graphs

To show the results of the proposed approach we used two large graphs that we have constructed. Both were processed using the procedure described in Section 2. The first graph is the collaboration graph between the institutions participating in European projects in the area of "New methods of work", where we had about 900 institutions (vertices). We have used a part of the graph generated in the   analyses of the database of research and development projects funded within information technology European program in years 2000-2005. The main items in the research project database were textual description of each project and the list of institutions participating in the project. There, the goal was to find various informative insights into the research project database, which would enable better understanding of the past dynamics and provide ground for better planning of the future research project programs.

The main emphasis was on the analysis of various aspects of research collaboration between different objects including institutions participating in the projects [4]. Figure 5 shows the results of drawing that graph using the approach proposed in this paper, zoomed-in around the German Fraunhofer Institute (the middle vertex in Figure 5)

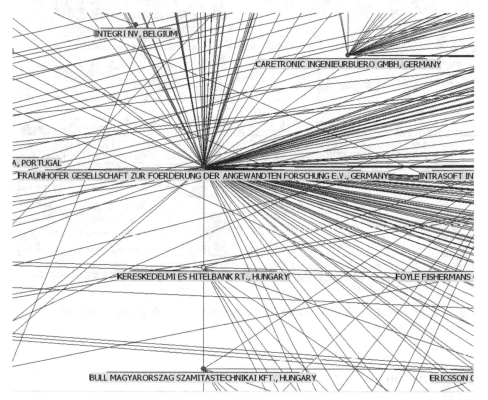

Fig. 5. The result of graph visualization (from the system user interface) on the collaboration of of institutions participating in European projects in the are of New methods of work. The program was instructed to zoom-in around one of the most connected institutions, the German Fraunhofer Institute

The second graph is the Amazon cross-sell graph (from year 2000) having about 10,000 books connected into a graph and capturing information about behavior of their customers. The two vertices in the graph (two books) are connected if there was a customer that both books. This kind of information can be further used by recommendation systems (via collaborative filtering) for making suggestion on potentially interesting books (cross-selling). Collaborative filtering in general is based on the assumption that similar users have similar preferences. It can detect relationships between items that are linked implicitly through the groups of users accessing them. Figure 6 shows output of our system on the Amazon cross-sell graph of books, where we have zoomed-in the part of the graph containing Java programming books. Titles placed together are such as, "Concurrent Programming in Java", "Concurrency: State

Model & Java Programming", "Practical Java Programming Language Guide", "Java in Practice: Design Symbols and Idioms for Effective Java".

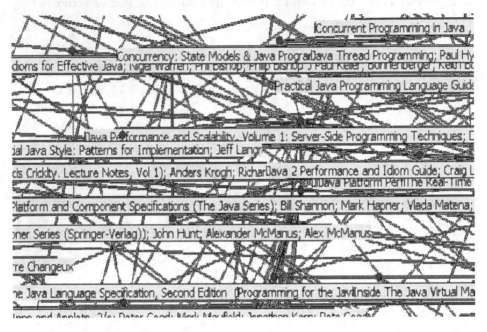

Fig. 6. The result of graph visualization on the Amazon cross-sell graph, zoom-in the part of Java programming books

4 Discussion

We have proposed a novel method for visualization of very large graphs, as we see graph as a fundamental data structure. We trust that many problems can be represented as graphs and that visualization of them can offer some insights in the data and particularly help in spotting local patterns. We expect that for finding local patterns, our approach can serve as one of the first steps for getting familiar with the data and in most cases should be combined with other approaches for further data analysis.

In our approach, we have applied data-analytic techniques for dealing with graphs. The original graph is transformed to a set of sparse vectors, one vector for each vertex. Then, hierarchical clustering is applied on the vectors and finally the hierarchy of clusters is used to assign coordinates in 2-D space to each graph vertex. We have provide results on two real-world example graphs, one from the domain of collaboration network between institutions and the other from the book cross-selling domain. In order to evaluate the results of our approach some kind of analytic and HCI evaluation of the method should be applied giving the same graphs to several systems and comparing their output as well as efficiency.

There are several directions that can be followed in the future work. One direction is on extending the proposed approach to deal with special types of graphs (e.g. power distributed), for instance, by include PageRank [10] weights. The other possible direction is in performing experiments to study the influence of different neighborhood weighting functions (see Equation 1).

References

1. Duda, R. O., Hart, P. E. and Stork, D. G.: Pattern Classification 2nd edition, WileyInterscience (2000)
2. Fayyad, U., Grinstein, G. G. and Wierse, A. (eds): Information Visualization in Data Mining and Knowledge Discovery, Morgan Kaufmann (2001)
3. Grobelnik, M., and Mladenić, D.: Efficient visualization of large text corpora. Proceedings of the seventh TELRI seminar. Dubrovnik, Croatia (2002)
4. Grobelnik, M., and Mladenić, D.: Analysis of a database of research projects using text mining and link analysis. In Mladenić, D., Lavrac, N., Bohanec, M., Moyle, S. (eds.), *Data mining and decision support : integration and collaboration*, (The Kluwer international series in engineering and computer science, SECS 745), pp. 157-166, Boston; Dordrecht; London: Kluwer Academic Publishers (2003)
5. Hand, D.J., Mannila, H., Smyth, P.: Principles of Data Mining (Adaptive Computation and Machine Learning), MIT Press (2001)
6. Hastie, T., Tibshirani, R. and Friedman, J. H.: The Elements of Statistical Learning: Data Mining, Inference, and Prediction, Springer Series in Statistics, Springer Verlag (2001)
7. Mitchell, T.M.: Machine Learning. The McGraw-Hill Companies, Inc. (1997)
8. Mutzel, P., Jünger, M., Leipert, S. (eds.): Graph Drawing : 9th International Symposium, GD-2001, Lecture Notes in Computer Science, Vol. 2265. Springer-Verlag, Berlin Heidelberg New York (2002)
9. North, S. (ed): Symposium on Graph Drawing GD'96, Lecture Notes in Computer Science, Vol. 1190. Springer-Verlag, Berlin Heidelberg New York (1997)
10. Page, L., Brin, S., Motwani, R., Winograd, T.: The PageRank citation ranking: bringing order to the web. Tech. Rept. SIDL-WP-1999-020, Stanford University, January (1998)
11. Robbins, K.S., Gorman, M.: Fast Visualization Methods for Comparing Dynamics: A Case Study in Combustion, Proceedings of the 11th IEEE Visualization 2000 Conference, IEEE Computer Society (2000)
12. Steinbach, M., Karypis, G., Kumar, V.: A comparison of document clustering techniques. In Proceedings of KDD Workshop on Text Mining, pp. 109–110 (2000)
13. Witten, I.H., Frank, E.: Data Mining: Practical Machine Learning Tools and Techniques with Java Implementations, Morgan Kaufmann (1999).

Features for Learning Local Patterns in Time-Stamped Data

Katharina Morik and Hanna Köpcke

Univ. Dortmund, Computer Science Department, LS VIII
morik@ls8.informatik.uni-dortmund.de
http://www-ai.cs.uni-dortmund.de

Abstract. Time-stamped data occur frequently in real-world databases. The goal of analysing time-stamped data is very often to find a small group of objects (customers, machine parts,...) which is important for the business at hand. In contrast, the majority of objects obey well-known rules and is not of interest for the analysis. In terms of a classification task, the small group means that there are very few positive examples and within them, there is some sort of a structure such that the small group differs significantly from the majority. We may consider such a learning task learning a local pattern.

Depending on the goal of the data analysis, different aspects of time are relevant, e.g., the particular date, the duration of a certain state, or the number of different states. From the given data, we may generate features that allow us to express the aspect of interest. Here, we investigate the aspect of state change and its representation for learning local patterns in time-stamped data. Besides a simple Boolean representation indicating a change, we use frequency features from information retrieval. We transfer Joachim's theory for text classification to our task and investigate its fit to local pattern learning. The approach has been implemented within the MiningMart system and was successfully applied to real-world insurance data.

1 Introduction

When designing a knowledge discovery application, the choice of the representation of examples and hypotheses is the most important issue. Choosing the right representation for hypotheses has been called "model selection". Learnability is a statement about a pair of example and hypotheses representation: we want to represent examples and hypotheses such that concepts of interest can be learned in at least polynomial time. This constrains the search for an appropriate representation on one side. The majority of known solutions to model selection deals with global models. What, if we are looking for local patterns? On the other side, the search is constrained by the given data. Transforming them into an appropriate representation is an effort, which we want to minimize. Moreover, not every example representation which is well suited for a learning algorithm can be constructed from given raw data. Figure 1 shows the chain of processes that

K. Morik et al. (Eds.): Local Pattern Detection, LNAI 3539, pp. 98–114, 2005.
© Springer-Verlag Berlin Heidelberg 2005

Fig. 1. Representation – from raw data to hypotheses

lead to the learning results. What are the characteristics of raw data that estimate whether we can construct an example representation for which, in turn, we know learnability results for a certain learning algorithm? Concerning the example representation, our standard approach to feature selection requires a subset of given features to separate the data according to the target concepts. What, if the learning task has to cope with an internal structure where attributes occurring in the target concept do occur in the remaining examples as well? It is our goal to develop guidelines for the choice of an appropriate example representation. We want to estimate *before* the transformation, whether it enlarges or shrinks the representation space and whether it favours a certain learning algorithm, or is appropriate for many of them. In terms of Figure 1, we are concerned with the question mark between raw data and the example language.

In this paper, we generalize the applicability of Thorsten Joachim's TCat model which gives clear learnability bounds for text classification using the support vector machine (SVM) [1]. The TCat model shares characteristics with local patterns (cf. Section 2). It refers to the bag-of-words representation of texts. It is straight forward to construct this example representation from texts. Now, we found that frequency features as used in the bag-of-words representation can effectively be constructed from time-stamped data, too. Moreover, we developed a heuristic that efficiently estimates for raw data, whether the transformation to frequency features will shrink or enlarge the data space, and whether it fits the distribution to which TCat models fit. For learning local patterns from time-stamped data we found that the representation enhanced the learning results of several algorithms, when compared to using the raw data.

The paper is organised as follows. We first state a working hypothesis on the notion of local patterns (Section 2). Then, we clarify the time aspects in time-stamped data (Section 3). For the aspect of state changes and their frequency, we show how to transform a relational database into one with frequency features for state change. We illustrate the procedure by real-world insurance data (Section 4). Then we can make use of the transformed data for learning a local pattern. We conducted several experiments with insurance data (Section 5.1). The good learning results are – for the SVM – explained by Joachim's theory. We describe the (transformed) insurance data by a TCat model (Section 6). We then focus on the the step from raw data to the TCat model. Section 6.1 characterizes the raw data with respect to the transformation and presents a heuristic which estimates the effect of the transformation. A program calculates a TCat model for given data set. Experiments with articifial data sets investigate the impact of the local patterns characteristics to TCat models and learning results (Section 7). This completes the chain from the time-stamped raw data to SVM learning of local patterns.

2   Local Patterns

Local patterns are not yet clearly defined. Common to all definitions is that we want to learn about rare events in a large collection. In other words, the distribution is skewed, offering few instances of the local pattern and very many instances of a global model. Local patterns occur with a low frequency (cf. the papers in [2]). Several authors have characterised local patterns as small regions with high deviations from a global model, e.g., [3], [4], [5]. Paul Cohen's definition of local patterns as "low frequency and low entropy" also relates the local pattern to the regular (global) frequency and entropy values. Moreover, it points at the internal stucture of local patterns or their examples. His learning task is to detect boundaries of episodes in a sequence. The comparison of frequency and entropy of one n-gram with those of all other n-grams of the same length delivers the standardized frequency and entropy [6]. Then, a local pattern is detected between two maximally (standardized) frequent n-grams and directly following an n-gram with highest (standardized) entropy. Episodes are complex patterns with an internal structure, which is represented here by the moving n-grams. Multimedia data (e.g., texts), time series, and DNA data are other instances of complex data types. Arno Siebes models the structure within complex data types using wavelets [4]. Transforming data into wavelets allows him to compute the similarity between examples which can then be used for further analysis. The wavelet transformation can be considered feature generation for structured data. We may find an internal structure also within a simple attribute-value representation. Interesting are the characteristics which Thorsten Joachims has found for texts in the bag of words representation [1]: The characteristic that several attribute values indicate the class membership has been called *a high level of redundancy*. The characteristic that instances do not (necessarily) share an attribute value being valid has been called *heterogenous use of terms*. In addition, we find that attribute values occurring frequently in one pattern do so in the remaining other observations, as well. We may call this an *overlap*. Redundancy, heterogenity, and overlap are characteristics of the internal structure of (text) instances.

As a working hypothesis, we end up with three characteristics of local patterns:

- Local patterns describe rare events. In other words, the distribution is skewed, offering few instances of the local pattern and very many instances of a global model.
- Given a dataset for which a global model can be determined, local patterns deviate significantly from the global model.
- Local patterns describe data with an internal structure. Redundancy, heterogenity, and overlap are aspects of internal structure.

3   Time-Stamped Data

Time-related data include time series (i.e. equidistant measurements of one process), episodes made of events from one or several processes, and time intervals which are related (e.g., an interval overlaps, precedes, or covers another interval). Time-stamped data refer to a calendar with its regularities. They can easily be transformed into a collection of events, can most often be transformed into time intervals, and sometimes into time series.

Time series are most often analysed with respect to a prediction task, but also trend and cycle recognition belong to the statistical standard (see for an overview [7,8]). Following the interest in very large databases, indexing of time series according to similarity has come into focus [9,10]. Clustering of time series is a related topic (cf. e.g., [11]) as is time series classification (cf. e.g., [12,13]). The abstraction of time series into sequences of events or time intervals approximates the time series piecewise by functions (so do [14,15,16]). Other segmentation methods are presented in [17].

Event sequences are investigated in order to predict events or to determine correlations of events [18,19], [20,21,22]. The approach of Frank Höppner abstracts time series to time intervals and uses the time relations of James Allen in order to learn episodes [23,24]. The underlying algorithm is one of learning frequent sets as is Apriori [25]. The resulting episodes are written as association rules. Other basic algorithms (e.g., regression trees) can be chosen as well [26], delivering logic rules with time annotations. Also inductive logic programming can be applied. Episodes are then written as a chain logic program, which expresses direct precedence by chaining unified variables and other time relations by additional predicates [27,28].

Time-stamped data have been investigated in-depth in [29]. They offer a framework for time granularities and specialised databases for temporal data. However, from a practical point of view, building up a temporal database before analysing the data ist too demanding for a knowledge discovery application. Hence, we prefer to transform the given data into an appropriate representation for data analysis. The given data are usually stored in a multirelational database where some attributes offer a time stamp of the same granularity (i.e., minute, hour, day, week, year). If time stamps in the stored data are equidistant, data actually are time series. We exclude this case, here. As was already stated above, time-stamped data can most often be transformed into events and time intervals. It is hard to select the appropriate representation [30]. In general, we have the following options:

Snapshot: We ignore the time information and reduce the data to the most current state. This state can be written as one or several events. It may well happen that such a snapshot already suffices for learning.

Events with time intervals: We aggregate time points to time intervals where attribute values are similar enough (segmentation). For nominal attributes, it is straight forward to construct time intervals from the start and end time of each attribute value. In addition, we might want to represent relations between the intervals. Learning algorithms which make good use of time information (episode learning) can then be applied.

Feature generation: Time aspects are encoded as regular attributes of the examples such that any learning algorithm can be applied. Simple encodings are seasons simply stated by flags and vectors, where the attributes summarize the history preceding a target event.

In this paper, we investigate feature generation for learning from time-stamped data. For each attribute, we represent whether the attribute changed at all over the recorded time span (Boolean), or how often it has changed (frequency feature). The latter allows us to link our results to the TCat model for text classification [1].

4   Using TF/IDF Features

Time-stamped data often describe the same object (customer, contract, engine...) by several rows in a database table, each for one of the object's states. Figure 2 illustrates this by an excerpt of an insurance contract database table, where each contract (VVID) is described by several attributes. Whenever an attribute's value has changed, a new row is added. The snapshot approach would

VVID	VVAENDNR	VVWIVON	VVWIBIS	VVAENDDAT	VVAENDART	...
16423	1	1946	1998	1946	1000	
16423	2	1998	1998	1998	27	
16423	3	1998	1998	1998	4	
16423	4	1998	1998	1998	54	
16423	5	1998	1998	1998	4	
16423	6	1998	9999	1998	61	
5016	1	1997	1999	1997	33	
5016	2	1999	2001	1999	33	
5016	3	2001	2001	2001	33	
5016	4	2001	2001	2001	33	
5016	5	2001	2002	2001	81	
5016	6	2002	9999	2002	94	
...						

history of a contract

history of another contract

Fig. 2. Excerpt from the contract table

just extract the most current row for a contract. The time interval approach would use the "begin" attribute (VVWIvon) and the "end" attribute (VVWIbis) and indicate the other attributes from a row as an event. A Boolean representation would just state whether an attribute had been changed over the lifetime of a contract, or not. This reduces the data space a lot. If we transform the raw data (about contracts) into a frequency representation, we possibly condense the data space in an appropriate way. We simply order the rows according to the data and count, how often an attribute's value changes, giving us the frequency count of that attribute. Figure 3 illustrates the procedure. However, we must

VVID	...	VVSTACD	VVPRFIN	VVPRZA	VVINKZWEI	VVBEG	VVEND	VVINKPRL	...
16423		4	1	2	2	1946	1998	295.29	
16423		4	1	2	2	1946	1998	295.29	
16423		4	5	2	0	1946	2028	0	
16423		5	3	2	0	1946	2028	0	
16423		4	1	2	2	1946	1998	295.29	
16423		5	3	2	0	1946	1998	0	

3	VVSTACD
4	VVPRFIN
0	VVPRZA
3	VVINKZWEI
0	VVBEG
2	VVEND
3	VVINKPRL

Fig. 3. Calculating the term frequency for the original attributes

exclude the frequencies of those changes that are common to all contracts, e.g. because of a change of law. The feature from statistical text representation formulates exactly this: term frequency and inverse document frequency (TFIDF) [31].

Term frequency here describes how often a particular attribute a_i of c_j, the contract or one of its components, has been changed.

$$tf(a_i, c_j) = \| \{x \in time\,points \mid a_i\,of\,c_j\,changed\} \|$$

The document frequency here corresponds to the number of contracts in which a_i has been changed. The set of all contracts is written C. The document frequency is just the number of contracts with a term frequency greater than 0.

$$df(a_i) = \| \{c_j \in C \mid a_i\,of\,c_j\,changed\} \|$$

Hence the adaptation of the TF/IDF text feature to contract data becomes for each contract c_j:

$$tfidf(a_i) = tf(a_i, c_j) log \frac{\| C \|}{df(a_i)}$$

5   Local Pattern Learning

Now that we have introduced frequency features for the aspect of change in time-stamped databases, we can bring together the local pattern learning and the time-stamped data. We first report on local pattern learning from a real-world database (Section 5.1). Then we present experiments with artificial data sets (Section 7).

5.1   Local Patterns in Insurance Data

In the course of enhanced customer relationship management, the Swiss Life insurance company investigated opportunities for direct marketing [32]. A more difficult task was to predict surrender in terms of a customer buying back his life insurance. We worked on knowledge discovery for the classification into early termination or continuation of policies. The task was clearly one of local pattern learning:

- Only 7.7% of the contracts end before their end date. Hence, the event to be predicted is rare.
- Internal studies at the insurance company found that for some attributes the likelihood of surrender differed significantly from the overall likelihood. The TCat model of the data (6) also clearly indicates this.
- Contract data have an internal structure.
  - First studies showed that frequent sets in the group of continued contract were frequent sets in the group of terminated contracts, as well [33] (overlap).
  - In each contract, there are several attributes indicating surrender or continuation (redundancy).
  - We also found that within the group of terminated contracts, there were those which do not share attributes (heterogenous use of terms).
  Hence, the internal structure of contracts shares characteristics with text data.

The given anonymous database consists of 12 tables with 15 relations between them. The tables contain information about 217,586 policies and 163,745 customers. If all records referring to the same policy and component (but at a different status at different times) are counted as one, there are 533,175 components described in the database. We selected 14 attributes from the original database. 13 of them were transformed as described above (Section 4). One of them is the reason for a change of a contract. There are 121 different reasons. We transformed these attribute values into binary attributes a. Thus we obtained 13+121=134 features describing changes of a contract. To calculate the TF/IDF values for these binary features we considered the history of each contract. For the 121 newly created features we counted how often they occurred within the mutations. Figure 4 shows how the calculation was done. We compared the learning results on this generated representation to those on the selected original data for different learning algorithms. We used 10-fold cross validation on a sample of 10,000 examples. In order to balance precision and recall, we used the F-measure:

$$F_\beta = \frac{(\beta^2 + 1)Prec(h)Rec(h)}{\beta^2 Prec(h) + Rec(h)} \tag{1}$$

where β indicates the relative weight between precision and recall. We have set $\beta = 1$, weighting precision and recall equally. Table 1 shows the results. For all algorithms, the frequency features are better suited than the original attributes.

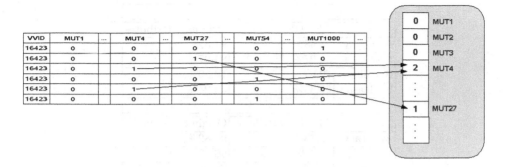

Fig. 4. Calculating the term frequency for the newly created features

6   Characterizing the Data by the TCat Model

The transformation into a frequency representation allows to model the data as TCat-concepts. TCat-concepts model text classification tasks such that their learnability can be proven [1].

Definition of TCat-concepts: *"The TCat-concept*

$$TCat([p_1 : n_1 : f_1], ..., [p_s : n_s : f_s])$$

describes a binary classification task with s sets of disjoint features. The i-th set includes f_i features. Each positive example contains p_i occurrences of features from the respective set, and each negative example contains n_i occurrences. The same feature can occur multiple times in one document." *[1]*

In order to describe the newly constructed data set in terms of TCat-concepts, we need to partition the feature space into disjoint sets of positive indicators, negative indicators and irrelevant features. For the insurance application, we selected features by their odds ratio. There are 2 high-frequency features that indicate positive contracts (surrender) and 3 high-frequency features indicating negative contracts (no surrender). Similarly, there are 3 (4) medium-frequency features that indicate positive (negative) contracts. In the low-frequency spectrum there are 19 positive indicators and 64 negative indicators. All other features are assumed to carry no information. Since the same feature can occur in p_i as well as in n_i, listing the features would not show the difference. The internal structure prohibits this. However, the number of occurrences clearly shows the significant difference between continued and early terminated contracts.

To abstract from the details of particular contracts, it is useful to define what a typical contract for this task looks like. An average contract has 8 features. For positive examples, on average 25% of the 8 features come from the set of the 2 high-frequency positive indicators while none of these features appear in an average negative contract. The relative occurrence frequencies for the other

	Apriori	
	TF/IDF attributes	Original attributes
Accuracy	93,48%	94,3%
Precision	56,07%	84,97%
Recall	72,8%	18,39%
F-Measure	63,35%	30,24%

	J4.8	
	TF/IDF attributes	Original attributes
Accuracy	99,88%	97,82%
Precision	98,64%	96,53%
Recall	99,8%	70,08%
F-Measure	99,22%	81,21%

	mySVM	
	TF/IDF attributes	Original attributes
Accuracy	99,71%	26,65%
Precision	97,06%	8,73%
Recall	98,86%	100%
F-Measure	97,95%	16,06%

	Naive Bayes	
	TF/IDF attributes	Original attributes
Accuracy	88,62%	87,44%
Precision	38,55%	32,08%
Recall	78,92%	77,72%
F-Measure	51,8%	45,41%

Table 1. Results comparing different learning algorithms and feature spaces

features are given in Table 2. Applying these percentages to the average number of features, this table can be directly translated into the following TCat-concept. Note, that p_i and n_i indicate frequencies. Hence, the second high-frequency set of features consists of three attributes, which occur one time in positive and three times in negative examples.

```
TCat ( [2 : 0 : 2], [1 : 4 : 3],      # high frequency
       [3 : 1 : 3], [0 : 1 : 4],      # medium frequency
       [1 : 0 : 19], [0 : 1 : 64],    # low frequency
       [1 : 1 : 39]                   # rest
     )
```

The learnability theorem of TCat-concepts [1] bounds the expected generalization error of an unbiased support vector machine after training on n examples by

$$\frac{R^2}{n+1} \frac{a + 2b + c}{ac - b^2} \tag{2}$$

	high frequency		medium frequency		low frequency		
	2 pos.	3 neg.	3 pos.	4 neg.	19 pos.	64 neg.	39 rest
pos. contract	25%	12.5%	37.5%	0%	12.5%	0%	12.5%
neg. contract	0%	50%	12.5%	12.5%	0%	12.5%	12.5%

Table 2. Composition of an average positive and an average negative contract

where R^2 is the maximum Euclidian length of any feature vector in the training data, and a, b, c are calculated from the TCat-concept description as follows:

$$a = \sum_{i=1}^{s} \frac{p_i^2}{f_i} \quad b = \sum_{i=1}^{s} \frac{p_i n_i}{f_i} \quad c = \sum_{i=1}^{s} \frac{n_i^2}{f_i}$$

$a = 5.41, b = 2.326, c = 5.952$ can be calculated directly from the data. The Euclidian length of the vectors remains to be determined. We want to see whether the data transformation condenses the data properly. The data space with the original 15 attributes could be such that each attribute is changed m times giving us $m\sqrt{15}$ – the largest case. The smallest case is that only one attribute is changed m times giving us the small data size of m. For texts, Zipf's law gives the approximation [34]: if one ranks the words by their frequency, the r-th most frequent words occur $\frac{1}{r}$ times the frequency of the most frequent words. We can apply this law for natural language to collections c of natural language texts. Experimental data suggests that Mandelbrot distributions [35]

$$TF_i = \frac{c}{(k+r)^\phi}$$

with parameters c, k and ϕ provide a better fit. For the contract data Figure 5 plots term frequency versus frequency rank. The line is an approximation with $k = -0.6687435$ and $\phi = 1.8$. We see that (as is true for text data) also the contract data can be shrinked by the frequency transformation.

$$R^2 = \sum_{r=1}^{d} \left(\frac{c}{(r+k)^2} \right)^2 \tag{3}$$

We bound $R^2 \leq 37$ according to the Mandelbrot distribution and come up with the bound of the expected error according to equation 3 of $\frac{37 \cdot 0.5978}{n+1}$, consequently after training on 1000 examples the model predicts an expected generalization error of less than 2.2%. It turns out that the transformed data sets can easily be separated by a support vector machine. Hence, the good learning results (0.6% error) are explained.

6.1 Characterizing the Raw Data

In order to ease the design process of knowledge discovery applications we should know before the transformation whether the data space will be condensed, or

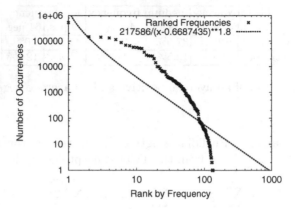

Fig. 5. Distribution of term frequencies in the contract data on a log log scale. The line is an approximation of the observed curve using a Mandelbrot distribution.

not. In other words, we want to measure the sparseness of data which can be estimated by the maximum Euclidian length of a vector. We order the original table with time stamps such that the states of the same individual (e.g., contract) are in succeeding rows. We consider each individual c a vector and calculate the frequency of changes for each of its n attributes $a_1...a_n$ in parallel "on the fly" We can then determine in one database scan the maximum value of the Euclidian length of a vector:

$$\hat{R} = max \left(\sqrt{\sum_{i=1}^{n} tf(a_i, c_j)^2} \right) \qquad (4)$$

If $\hat{R} \leq \frac{m\sqrt{n}}{2}$ where m is the maximum frequency, the data will be condensed and learning will be fast. In the insurance case $n = 14$ and $m = 15$ so that $\hat{R} = 22,913$ which is in fact less than $\frac{15\sqrt{14}}{2} = 28.6$.

Of course, the heuristic does not tell anything about the learnability within this representation. What we control with this heuristic is the transformation into the example representation.

7   Experiments with Artificial Data

In order to abstract away from the real-world application, we conducted experiments with artificial datasets. We created 10,000 examples each with 100 attributes in both, binary and TF/IDF representation. The MYSVM was run with a 10-fold cross validation.

7.1   Frequency of Changes Vs. Particular Changes

Datasets were generated according to two target concepts:

1. The change of particular attributes determined the classification.
2. The number of changes determined the target concept.

Of course, what should happen, is that the TF/IDF representation is best suited for the second target concept.

We wanted to check, whether the TF/IDF representation makes the dataset robust with respect to skewedness and local structure. In other words, we tested the characteristic of local patterns.

We systematically varied the skewedness of the data, positive examples being 50% (not skewed), 25%, 12.5%, or 6.25% (skewed) of the data.

The first concept could perfectly be learned from the binary representation, being robust with respect to skewedness (100% F-measure). Clearly, no internal structure is preventing here the selection of a feature. The TF/IDF representation is little less perfect (95.03% F-measure for 6.25% positive examples). See Figure 6 for details. The heuristic states that the data size becomes little. As is shown in Figure 6, the second concept could not be learned from the binary representation, when the distribution becomes skewed. The default is too dominant, hence the recall comes to zero. For the TF/IDF representation, the learning results degraded gracefully when the distribution became more skewed from 93.17% to 88.98% F-measure. The heuristic dissuades from the transformation in all cases, although the binary representation is only superior learning the first concept. Indeed, as already stated, the larger data space is necessary for the skewed second concept. The heuristic does not inform us about learnability. Hence, when learning the second concept from skewed data fails using the binary representation, the transformation is tried, anyway.

Varying the sparseness of the data from 50 attributes being changed over 25 to 5 attributes being changed, we found again that using the binary representation the first concept could be learned perfectly. In contrast, the second concept could only be learned successfully using the TF/IDF representation. See Figure 6 for details.

7.2   Internal Structures

We also varied the local structure within the artificial data. We used the TCat model to generate datasets, varying p_i and n_i within f_i. Note, that for a group of f_i (high frequency, medium frequency, or low frequency) p_i of them can be arbitrarily chosen. Hence, it could happen that $\frac{f_i}{p_i}$ individuals do not share any attribute. This fraction indicates how heterogenous the use of terms is. We varied the heterogenous use of terms from 4 individuals which could be completely different but be in the same class (little heterogenity) to 20 individuals being disjoint but in the same class. If we keep the sparseness throughout all experiments being 20 from 100 attributes given, we automatically vary the redundancy from 0.5 in the little heterogenous case to 0.1 in the extremely heterogenous case. The redundancy can be expressed by $\frac{p_i}{f_i}$ or $\frac{n_i}{f_i}$ for indicative attributes (here:medium and low frequency attributes). Tabelle 3 gives an overview of the TCat models used for generating datasets. Figure 6 shows the achieved learning results.

	Not heterogenous, redundant, no overlap	Little heterogeneous, redundant, little overlap	Medium heterogenous, little redundant, no overlap
heterogenity	$\frac{20}{5}=4$	$\frac{20}{4}=5$	$\frac{20}{2}=10$
redundancy	$\frac{10}{20}=0.5$	$\frac{10}{20}=0.5$	$\frac{4}{20}=0.2$
TCat model	[10 : 10 : 20], [5 : 0 : 20], [0 : 5 : 20], [5 : 0 : 20], [0 : 5 : 20],	[10 : 10 : 20], [4 : 1 : 20], [1 : 4 : 20], [4 : 1 : 20], [1 : 4 : 20],	[16 : 16 : 20] [2 : 0 : 20], [0 : 2 : 20], [2 : 0 : 20], [0 : 2 : 20],
error bound	1.33%	3.3%	3%
	High heterogenous, little redundant, no overlap	Medium heterogenous, redundant, high overlap	
heterogenity	$\frac{20}{1}=20$	$\frac{20}{3}=6.6$	
redundancy	$\frac{2}{20}=0.1$	$\frac{10}{20}=0.5$	
TCat model	[18 : 18 : 20] [1 :0 : 20], [0 : 1 : 20], [1 : 0 : 20], [0 : 1 : 20],	[10 : 10 : 20] [3 : 2 : 20], [2 : 3 : 20], [3 : 2 : 20], [2 : 3 : 20],	
error bound	72.4%	28%	

Table 3. TCat models used for generating datasets

The TCat model tells that the learning results should decrease gracefully when increasing heterogenous use of terms or overlap. The actual learning results show that only the overlap decreases the learning result and only for the binary representation. The TF/IDF representation is robust to the variation of both, heterogenous use of terms and overlap. This means that the TF/IDF representation is particularly appropriate for the internal structure within local patterns.

8   Conclusion

The design of a knowledge discovery application is supported by learnability results as soon as the appropriate example representation has been found. The design support missing was a principled approach to when to generate which features for a given dataset. The transformation of given (raw) data to the example representation is a matter of feature generation and selection [36,37,38]. The automatic selection of features becomes difficult whenever no proper subset of features distinguishes positive from negative examples. This is particularly the case, if the target concept has an internal structure in terms of the heterogenous use, redundancy, and overlap of attributes. If, in addition, the distribution offers a majority of negative and very few positive examples, we are confronted

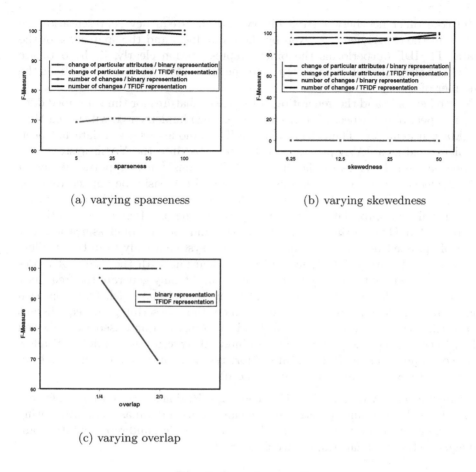

(a) varying sparseness

(b) varying skewedness

(c) varying overlap

Fig. 6. Learning results

with local pattern learning. The TCat model covers such an internal structure, is robust with respect to skewed distributions, and offers learnability results for the SVM. Hence, if we form a TCat model for our data, we can estimate how well learning will succeed. However, the TCat model is based on frequency features. Calculating frequency features for texts is a standard approach. Now, we have shown how time-stamped data can be transformed into frequency features. Moreover, experiments with artifical datasets indicate that

- the TF/IDF representation is superior to the binary one if attributes indicative within the positive and negative example sets overlap.
- The TF/IDF representation outperforms the binary one in case of sparse data.
- the TF/IDF representation is robust with respect to skewed data
- the TF/IDF representation is robust with respect to redundancy or heterogenous use of terms.

The true difference between the binary and the frequency representation we found in the experiments is given by the skewedness and the sparseness of the data. TF/IDF outperforms the binary representation clearly, if the data are sparse or the distribution is skewed and the concept to be learned is about the number of changes.

The heuristic and the generation of frequency features for time-stamped data with respect to the aspect of state change has been implemented within the MiningMart system [39]. The calculation of a TCat model for given data has been implemented as a JAVA program to be integrated into the YALE system [38], which has run the cross validation and MYSVM runs. This supports the design of knowledge discovery applications. The method can easily be applied to other time-stamped datasets. If a general model can be found, the binary representation will be appropriate. For a local pattern, the TF/IDF representation is better suited. Given a time-stamped database and the task of classification (not that of episode learning), one can now proceed systematically from the smallest to successively larger data spaces. First, try learning with the snapshot representation, then try the binary representation, and finally generate the frequency features and calculate the error bound using the TCat model. Our approach only covers a certain type of raw data, namely those describing non-equidistant state changes. It focuses on finding local patterns, characterised by a skewed distribution, separable patterns, and an internal structure which does not allow to select a proposer subset of features. More research on the transformation from raw data to example representations is needed.

Acknowledgment We thank Jörg-Uwe Kietz and Regina Zücker for their information about the insurance practice and the anonymised database. For stimulating discussions on the support vector machine, mathematics, and representation languages we thank Stefan Rüping wholeheartedly.

References

1. Joachims, T.: Learning to Classify Text using Support Vector Machines. Volume 668 of Kluwer International Series in Engineering and Computer Science. Kluwer (2002)
2. Hand, D., Bolton, R., Adams, N.: Determining hit rate in pattern search. In Hand, D., Adams, N., Bolton, R., eds.: Pattern Detection and Discovery. Springer (2002)
3. Hand, D.: Pattern detection and discovery. In Hand, D., Adams, N., Bolton, R., eds.: Pattern Detection and Discovery. Springer (2002)
4. Siebes, A., Struzik, Z.: Complex data mining using patterns. In Hand, D., Adams, N., Bolton, R., eds.: Pattern Detection and Discovery. Springer (2002)
5. Morik, K.: Detecting interesting instances. In Hand, D.J., Adams, N.M., Bolton, R.J., eds.: Proceedings of the ESF Exploratory Workshop on Pattern Detection and Discovery. Volume 2447 of LNAI., Berlin, Springer Verlag (2002) 13–23
6. Paul Cohen, Brent Heeringa, Niall M. Adams: An unsupervised algorithm for segmenting categorical timeseries into episodes. In Hand, D.J., Adams, N.M., Bolton, R.J., eds.: Pattern Detection and Discovery. Volume 2447 of Lecture notes in computer science., London, UK, ESF Exploratory Workshop, Springer (2002) 1–12

7. Box, G.E.P., Jenkins, G.M., Reinsel, G.C.: Time Series Analysis. Forecasting and Control. Third edn. Prentice Hall, Englewood Cliffs (1994)
8. Schlittgen, R., Streitberg, B.H.J.: Zeitreihenanalyse. 9. edn. Oldenburg (2001)
9. Keogh, E., Pazzani, M.: Scaling up dynamic time warping for datamining applications. In: Proceedings of the 6th ACM SIGKDD International Conference on Knowledge Discovery and Data Mining, ACM Press (2000) 285–289
10. Agrawal, R., Faloutsos, C., Swami, A.: Efficient similarity search in sequence databases. In: Proceedings of the 4th International Conference on Foundations of Data Organization and Algorithms. Volume 730., Springer (1993) 69–84
11. Oates, T., Firoiu, L., Cohen, P.R.: Using dynamic time warping to bootstrap hmm-based clustering of time series. In: Sequence Learning ? Paradigms, Algorithms, and Applications. Volume 1828 of Lecture Notes in Computer Science. Springer Verlag (2001) 35?–52
12. Geurts, P.: Pattern extraction for time series classification. In: Pro-ceedings of the 5th European Conference on the Principles of Data Mining and Knowledge Discovery. Volume 2168 of Lecture Notes in Computer Science., Springer (2001) 115–127
13. Lausen, G., Savnik, I., Dougarjapov, A.: Msts: A system for mining sets of time series. In: Proceedings of the 4th European Conference on the Principles of Data Mining and Knowledge Discovery. Volume 1910 of Lecture Notes in Computer Science., Springer Verlag (2000) 289–298
14. Das, G., Lin, K.I., Mannila, H., Renganathan, G., Smyth, P.: Rule Discovery from Time Series. In Agrawal, R., Stolorz, P.E., Piatetsky-Shapiro, G., eds.: Proceedings of the Fourth International Conference on Knowledge Discovery and Data Mining (KDD-98), New York City, AAAI Press (1998) 16 – 22
15. Guralnik, V., Srivastava, J.: Event detection from time series data. In: Proceedings of the fifth ACM SIGKDD international conference on Knowledge discovery and data mining, San Diego, USA (1999) 33 – 42
16. Morik, K., Wessel, S.: Incremental signal to symbol processing. In Morik, K., Kaiser, M., Klingspor, V., eds.: Making Robots Smarter – Combining Sensing and Action through Robot Learning. Kluwer Academic Publ. (1999) 185 –198
17. Salatian, A., Hunter, J.: Deriving trends in historical and real-time continuously sampled medical data. Journal of Intelligent Information Systems 13 (1999) 47–71
18. Agrawal, R., Psaila, G., Wimmers, E.L., Zaït, M.: Querying shapes of histories. In: Proceedings of 21st International Conference on Very Large Data Bases, Morgan Kaufmann (1995) 502–514
19. Domeniconi, C., shing Perng, C., Vilalta, R., Ma, S.: A classification approach for prediction of target events in temporal sequences. In Elomaa, T., Mannoila, H., Toivonen, H., eds.: Principles of Data Mining and Knowledge Discovery. Lecture Notes in Artificial Intelligence, Springer (2002)
20. Blockeel, H., Fürnkranz, J., Prskawetz, A., Billari, F.: Detecting temporal change in event sequences: An application to demographic data. In De Raedt, L., Siebes, A., eds.: Proceedings of the 5th European Conference on the Principles of Data Mining and Knowledge Discovery. Volume 2168 of Lecture Notes in Computer Science., Springer (2001) 29–41
21. Mannila, H., Toivonen, H., Verkamo, A.: Discovering frequent episode in sequences. In: Procs. of the 1st Int. Conf. on Knowledge Discovery in Databases and Data Mining, AAAI Press (1995)
22. Mannila, H., Toivonen, H., Verkamo, A.: Discovery of frequent episodes in event sequences. Data Mining and Knowledge Discovery 1 (1997) 259–290

23. Höppner, F.: Discovery of Core Episodes from Sequences. In Hand, D.J., Adams, N.M., Bolton, R.J., eds.: Pattern Detection and Discovery. Volume 2447 of Lecture notes in computer science., London, UK, ESF Exploratory Workshop, Springer (2002) 1–12

24. Allen, J.F.: Towards a general theory of action and time. Artificial Intelligence 23 (1984) 123–154

25. Agrawal, R., Imielinski, T., Swami, A.: Database mining: A performance perspektive. IEEE Transactions on Knowledge and Data Engineering 5 (1993) 914–925

26. Nunez, M.: Learning patterns of behavior by observing system events. In: Procs. of the European Conference on Machine Learning. Lecture notes in Artificial Intelligence, Springer (2000)

27. Klingspor, V., Morik, K.: Learning understandable concepts for robot navigation. In Morik, K., Klingspor, V., Kaiser, M., eds.: Making Robots Smarter – Combining Sensing and Action through Robot Learning. Kluwer (1999)

28. Rieger, A.D.: Program Optimization for Temporal Reasoning within a Logic Programming Framework. PhD thesis, Universität Dortmund, Dortmund, Germany (1998)

29. Bettini, C., Jajodia, S., Wang, S.: Time Granularities in Databases, Data Mining, and Temporal Reasoning. Springer (2000)

30. Morik, K.: The representation race - preprocessing for handling time phenomena. In de Mántaras, R.L., Plaza, E., eds.: Proceedings of the European Conference on Machine Learning 2000 (ECML 2000). Volume 1810 of Lecture Notes in Artificial Intelligence., Berlin, Heidelberg, New York, Springer Verlag Berlin (2000)

31. Salton, G., Buckley, C.: Term weighting approaches in automatic text retrieval. Information Processing and Management 24 (1988) 513–523

32. Kietz, J.U., Vaduva, A., Zücker, R.: Mining Mart: Combining Case-Based-Reasoning and Multi-Strategy Learning into a Framework to reuse KDD-Application. In Michalki, R., Brazdil, P., eds.: Proceedings of the fifth International Workshop on Multistrategy Learning (MSL2000), Guimares, Portugal (2000)

33. Fisseler, J.: Anwendung eines Data Mining-Verfahrens auf Versicherungsdaten. Master's thesis, Fachbereich Informatik, Universität Dortmund (2003)

34. G.K.Zipf: Human Behavior and the Principle of Least Effort: An Introduction to Human Ecology. Addison-Wesley (1949)

35. Mandelbrot, B.: A note on a class of skew distribution functions: Analysis and critique of a paper by H.A.Simon. Informationi and Control 2 (1959) 90 – 99

36. Kohavi, R., John, G.H.: Wrappers for feature subset selection. Artificial Intelligence 97 (1997) 273–324

37. Liu, H., Motoda, H.: Feature Extraction, Construction, and Selection: A Data Mining Perspective. Kluwer (1998)

38. Ritthoff, O., Klinkenberg, R., Fischer, S., Mierswa, I.: A hybrid approach to feature selection and generation using an evolutionary algorithm. Technical Report CI-127/02, Collaborative Research Center 531, University of Dortmund, Dortmund, Germany (2002) ISSN 1433-3325.

39. Morik, K., Scholz, M.: The MiningMart Approach to Knowledge Discovery in Databases. In Zhong, N., Liu, J., eds.: Intelligent Technologies for Information Analysis. Springer (2004)

Boolean Property Encoding for Local Set Pattern Discovery: An Application to Gene Expression Data Analysis

Ruggero G. Pensa and Jean-François Boulicaut

INSA Lyon
LIRIS CNRS UMR 5205
F-69621 Villeurbanne cedex, France
{Ruggero.Pensa,Jean-Francois.Boulicaut}@insa-lyon.fr

Abstract. In the domain of gene expression data analysis, several re-searchers have recently emphasized the promising application of local pattern (e.g., association rules, closed sets) discovery techniques from boolean matrices that encode gene properties. Detecting local patterns by means of complete constraint-based mining techniques turns to be an important complementary approach or invaluable counterpart to heuristic global model mining. To take the most from local set pattern mining approaches, a needed step concerns gene expression property encoding (e.g., over-expression). The impact of this preprocessing phase on both the quantity and the quality of the extracted patterns is crucial. In this paper, we study the impact of discretization techniques by a sound comparison between the dendrograms, i.e., trees that are generated by a hierarchical clustering algorithm on raw numerical expression data and its various derived boolean matrices. Thanks to a new similarity measure, we can select the boolean property encoding technique which preserves similarity structures holding in the raw data. The discussion relies on several experimental results for three gene expression data sets. We believe our framework is an interesting direction of work for the many application domains in which (a) local set patterns have been proved useful, and (b) Boolean properties have to be derived from raw numerical data.

1 Introduction

This volume is dedicated to local pattern detection. It has been motivated by the need for a better characterization of what is local pattern detection and what are the main research challenges in this area. We contribute to this objective by considering the exciting application domain of *transcription module discovery* from gene expression data. In this molecular biology context, the goal is to identify sets of genes which seem to be co-regulated, associated with the sets of biological situation which seems to trigger the co-regulation.

The state-of-the-art is that global patterns like partitions can provide some useful information and suggest some of the transcription modules. We are how-ever interested by the intrinsic limitations of these approaches, e.g., their heuris-tic nature or the lack of unexpectedness of the findings. We strongly believe

K. Morik et al. (Eds.): Local Pattern Detection, LNAI 3539, pp. 115–134, 2005.

that complete extractions of local patterns which satisfy a given conjunction of constraints (e.g., a minimal frequency constraint or a maximality constraint) are an invaluable and complementary approach to suggest unexpected but relevant patterns, i.e., putative transcription modules.

Let us now introduce the application domain and our contribution. Thanks to a huge research effort and technological breakthroughs, one of the challenges for molecular biologists is to discover knowledge from data generated at very high throughput. For instance, different techniques (including microarray [1] and SAGE [2]) enable to study the simultaneous expression of (tens of) thousands of genes in various biological situations. The data generated by those experiments can be seen as expression matrices in which the expression level of genes (rows) is recorded in various biological situations (columns). A toy example of some microarray data is the matrix in Tab. 1a.

	1	2	3	4	5
a	-1	6	0	12	9
b	3	-2	3	-3	1
c	0	5	-1	6	6
d	4	-1	2	-2	-1
e	-3	9	1	10	6
f	5	-3	3	-6	0
g	4	-4	3	-7	0
h	-2	2	-2	8	5

(a)

	1	2	3	4	5
a	0	1	0	1	1
b	1	0	1	0	1
c	0	1	0	1	1
d	1	0	1	0	0
e	0	1	0	1	1
f	1	0	1	0	1
g	1	0	1	0	1
h	0	0	0	1	1

(b)

	1	2	3	4	5
a	0	0	0	1	0
b	1	0	1	0	0
c	0	0	0	1	1
d	1	0	0	0	0
e	0	0	0	1	0
f	1	0	0	0	0
g	1	0	0	0	0
h	0	0	0	1	0

(c)

Table 1. A gene expression matrix (a) with two derived boolean matrices (b and c)

Once large gene expression datasets are available, biologists have to drop the traditional one-to-one approach to gene expression data analysis and crucially need for Knowledge Discovery in Databases techniques (KDD). Among the classical KDD approaches, classification techniques (i.e., learning a classifier from data which, for example, can predict a cancer diagnosis according to individual gene expression profiles) have been intensively studied (see, e.g., [3] for a collection of recent contributions). In this paper, we do not consider such problems. We are interested in descriptive techniques which provides either global patterns like partitions (clustering) or local patterns like co-regulated sets of genes and/or sets of situations.

The use of hierarchical clustering (see, e.g., [4]) is indeed quite popular among practitioners. Genes are grouped together according to similar expression profiles. The same can be done on biological situations. Thanks to the appreciated vizualization component introduced with [4], biologists can identify some putative transcription modules. Practitioners do not use only hierarchical clustering but also most of the classical clustering techniques. A common characteristic of these techniques is that global patterns like partitions are extracted by means of a heuristic search. They provide "global pictures" of similarity structures. Not only the heuristic nature can lead to different results for different experiments but also, the fact we get global patterns, i.e., which hold in the whole data, leads to rather expected findings. Our thesis is that unexpected patterns are a priori interesting and that they are typically local ones, i.e., they hold in only a part of the data. Therefore, looking for collections of local patterns in gene expression data appears as a promising and complementary approach. The last 5 years, a major research sub-domain in data mining has concerned the design of efficient and complete constraint-based mining tools on boolean data, also called trans-actional data by some authors. The completeness assumption means that every pattern from the pattern language which satisfies the defined constraints has to be returned (e.g., every frequent set, every closed set, every frequent and closed set which does not contain a given item). In general, and this is the case for our work, non heuristic methods are used.

To apply these techniques for gene expression data analysis, we have to encode boolean gene expression properties, e.g., over-expression, strong variation, co-regulation. Tab. 1b and Tab. 1c are two data sets derived from the toy microarray data from Tab. 1a. Once such boolean data sets are available, it is possible to look for putative synexpression groups (see [5]) by computing the popular frequent sets (frequent sets of situations in a matrix $Genes \times Situations$ and frequent sets of genes in its transposition). Given the number of genes, we can alternatively compute condensed representations of the frequent sets, e.g., the frequent closed sets [6,7,8]. Deriving association rules from synexpression groups has been studied as well [9,10]. Furthermore, putative transcription modules can be provided by computing the so-called formal concepts (see, e.g., [11,12,13]). Also, constraint-based mining of concepts has been considered [14,15]. Notice that the collection of every formal concept which can be extracted from large real gene expression matrices can be considered as a collection of overlapping clusters on either the genes or the situations. The global picture is not there but every locally strong association (associated closed sets, see Section 3) has been captured.

So far, very few studies have concerned the quality of gene expression property encoding, i.e., a kind of feature construction phase. This is a critical step because its impact on both the quantity and the quality of the extracted patterns is crucial.

If S denotes the set of biological situations and P denotes the set of genes, the expression properties can be encoded into $r \subseteq P \times S$. $(g_i, o_j) \in r$ denotes that gene i has the encoded expression property in situation j. Different expression properties might be considered. Without loss of generality, we consider that only one expression property is encoded for each gene, which means that we can talk indifferently of genes or gene expression properties. Generally, encoding is performed according to some discretization operators that, given user-defined parameters, transform each numerical value from raw gene expression data into one boolean value per gene property. Many operators can be used that typically compute thresholds from which it is possible to decide wether the true or the false value must be assigned. For instance, in Tab. 1b, an over-expression property has been encoded and, e.g., Genes a, c, and e are over-expressed together in Situations 2, 4 and 5.

In [16], we have proposed a method which supports the choice for a discretization technique and an informed decision about its parameters. The idea was to study the impact of discretization by a sound comparison between the dendrograms (i.e., binary trees) that are generated by the same hierarchical clustering algorithm applied to both the raw expression data and various derived boolean matrices. This paper is a significant extension of [16]. The framework has been revisited and the experimental validation have been considerably extended.

In Section 2, we refine the similarity measure introduced in [16]. It is level independent, and it depends for each node on its subtree structure. It can be applied on gene and/or situation dendrograms and we introduce an aggregated measure for considering both simultaneously. Section 3 is dedicated to the use of this similarity measure on three real gene expression data sets in order to select an adequate discretization technique. The robustness of the approach is also emphasized by an a posteriori analysis of the extracted patterns in the various boolean contexts. For this purpose, we adapt the similarity measure between collections of patterns introduced in [17]. In Section 4, we study further the robustness of our approach by comparing several clustering results in the raw data. Section 5 is a short conclusion.

2   Boolean Encoding Assessment

2.1   Comparing Binary Trees

The problem of tree comparison has motivated a lot of research. Designing similarity measures between trees is difficult because it has to be defined according to the semantics of trees and similarities which are generally application domain dependant. For instance, considering the analysis of phylogenies, distance measures between both rooted and unrooted trees have been designed to compare different phylogenetic trees concerning the same set of individuals (e.g., different species of animals having a common ancestor). Various distance metrics between trees have been proposed. The nni (nearest neighbor interchange) and the mast (maximum agreement subtree) are two of the most used metrics. nni has been introduced independently in [18] and [19] and its NP-completeness has been recently proved [20,21]. mast has been proposed in [22], and [23] describes an efficient algorithm for computing this metrics on binary trees. These two approaches are tailored for the problem of comparing phylogenies where the goal is to measure some degree of isomorphism between two dendrograms representing the same species of biological organisms.

In our data mining problem, we have sets of objects (vectors of expression values for genes in various biological situations), that we want to process with a hierarchical clustering algorithm. Depending on the different discretization operations on raw expression data, a same clustering algorithm working on encoded

boolean gene expression data can return (very) different results. We are looking for a method that supports the comparison of these various gene and/or
situation dendrograms obtained on boolean data w.r.t. the common reference
dendrogram that has been computed from the raw data. We need to measure
both the degree of similarity of their structures and the similarity between the
contents of their associated collections of clusters. We introduced in [16] a simple
measure which is also easy to compute. Intuitively, it depends on the number of
matching nodes between the two trees we have to compare.

2.2   Definition of Similarity Scores

Let $\mathcal{O} = \{o_1, \dots, o_n\}$ denote a set of n objects. Let T denote a binary tree built
on $\mathcal{O}$. Let $\mathcal{L} = \{l_1, \dots, l_n\}$ denote the set of n leaves of T associated to $\mathcal{O}$ for
which, $\forall i \in [1 \dots n]$, $l_i \equiv o_i$. Let $\mathcal{B} = \{b_1 \dots b_{n-1}\}$ denote the set of the $n - 1$
nodes of T generated by a hierarchical clustering algorithm starting from $\mathcal{L}$. By
construction, we consider $b_{n-1} = r$, where r denotes the root of T. We define
the two sets:

$$\delta\,(b_i) = \{b_j \in \mathcal{B} \mid b_j \text{ is a descendent of } b_i\},$$

$$\tau\,(b_i) = \{l_j \in \mathcal{L} \mid l_j \text{ is a descendent of } b_i\}.$$

An example of a tree for the genes from Tab. 1a is given in Fig. 1. Here,
$\tau\,(b_3) = \{b, d, f, g\}$ and $\delta\,(b_3) = \{b_1, b_2\}$.

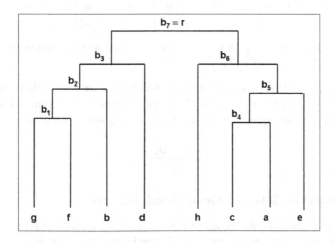

Fig. 1. An example of binary tree

We want to measure the similarity between a tree T and a reference tree T_{ref}
built on the same set of objects $\mathcal{O}$. For each node b_i of T, we define the following
score (denoted S_B and called **BScore**):

$$S_B\left(b_i, T_{ref}\right) = \sum_{b_j \in \delta(b_i)} a_j$$

$$a_j = \begin{cases} \frac{1}{|\tau(b_j)|}, & if \\ \quad \exists b_k \in T_{ref} \mid \tau\left(b_j\right) = \tau\left(b_k\right) \\ 0, & otherwise \end{cases} \tag{1}$$

In other terms, for a node b in T, its score depends both on the number of its matching nodes in T_{ref} ($b_k \in T_{ref}$ is a matching node for b if $\tau\left(b\right) = \tau\left(b_k\right)$) and $|\tau(b)|$. To obtain the similarity score of T w.r.t. T_{ref} (denoted S_T and called **TScore**), we consider the **BScore** value on the root, i.e.:

$$S_T\left(T, T_{ref}\right) = S_B\left(r, T_{ref}\right) \tag{2}$$

As usually, it is interesting to normalize the measure to get a score between 0 (for a tree which is totally different from the reference) and 1 (for a tree which is equal to the reference). For the **TScore** measure, since its maximal value depends on the tree morphology, we can normalize by $S_T\left(T_{ref}, T_{ref}\right)$:

$$\overline{S_T}\left(T, T_{ref}\right) = \frac{S_T\left(T, T_{ref}\right)}{S_T\left(T_{ref}, T_{ref}\right)} \tag{3}$$

$\overline{S_T}\left(T, T_{ref}\right) = 0$ means that T is totally different from T_{ref}, i.e., there are no matching node between T and T_{ref}. Indeed, $\overline{S_T}\left(T, T_{ref}\right) = 1$ means that T is totally similar to T_{ref}, i.e., every node in T matches with a node in T_{ref}. Given two trees T_1 and T_2 and a reference T_{ref}, if $\overline{S_T}\left(T_1, T_{ref}\right) < \overline{S_T}\left(T_2, T_{ref}\right)$, then T_2 is said to be more similar to T_{ref} than T_1 according to **TScore**.

An important property (missing from [16]) is the following:

Property 1. The measure 1 is asymmetric, i.e. given a reference tree T_{ref}, $\exists T$ such that $\overline{S_T}\left(T, T_{ref}\right) \neq \overline{S_T}\left(T_{ref}, T\right)$.

As a consequence of this property, such a measure makes sense when one wants to compare different binary trees with the same reference. If a symmetric measure is needed, one can consider the mean of the two possible measures for a couple of trees:

$$\frac{\overline{S_T}\left(T_1, T_2\right) + \overline{S_T}\left(T_2, T_1\right)}{2}.$$

2.3   Comparison Between Gene Dendrograms

Tab. 1a is a toy example of a gene expression matrix. Each row represents a gene vector, and each column represents a biological sample vector. Each cell contains an expression value for a given gene and a given sample. In this example, we have $\mathcal{O} = \{a, b, c, d, e, f, g, h\}$. A hierarchical clustering using the Pearson's correlation coefficient and the average linkage method (see, e.g., [4]) on the data from Tab. 1a leads to the dendrogram in Fig. 1.

Assume now that we discretize the expression matrix by applying two different methods used for over-expression encoding [9]. The first one, the so-called

"Mid-Ranged" method, considers the mean between the maximal and minimal values for each gene vector. Values which are greater than the average value are set to 1, 0 otherwise (Tab. 1b). A second method, the so-called "Max - X% Max" method, takes into account the maximal value for each gene vector. Values that are greater than $(100 - X)\%$ of the maximal value are set to 1, 0 otherwise. We set X to 10 deriving the matrix in Tab. 1c.

Assume now that we use the same clustering algorithm on the two derived boolean data sets. The resulting dendrograms are shown in Fig. 2. Fig. 2a (resp. Fig. 2b) represents the gene dendrogram obtained by clustering the boolean matrix in Tab. 1b (resp. Tab. 1c).

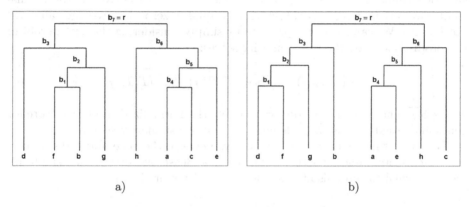

a)   b)

Fig. 2. Gene trees built on two differently discretized matrices

We can now use the similarity score and decide which discretization is better for this gene expression data set, i.e., the one for which $\overline{S_T}(T, T_{ref})$ has the largest value. The common reference (T_{ref}) is the tree in Fig. 1. Let T_a and T_b denote the trees in Fig. 2a and 2b respectively. Using Equation 3, we obtain:

$$\overline{S_T}(T_a, T_{ref}) = 0.77 \qquad \overline{S_T}(T_b, T_{ref}) = 0.23.$$

Since $\overline{S_T}(T_a, T_{ref}) > \overline{S_T}(T_b, T_{ref})$, the first discretization method is considered better for this data set w.r.t. the performed hierarchical clustering. In fact, in T_a, only node b_1 does not match (i.e., it does not share the same set of leaves) with any node in T_{ref}, while in T_b, there are only two nodes (b_3 and b_6) that match with some nodes in T_{ref}.

The same process can be applied to situation dendrograms by considering now that the objects are the situations. In practice, we perform both processes to support the choice of a discretization technique as illustrated in the next section.

2.4   Average Similarity Score

When we compare both situation and gene trees, we have different results for each comparison. According to our practice of gene expression data analysis, we often have thousands genes and a few tens or hundreds of situations. It means that, the similarity scores computed for situations tree are usually greater than those computed for gene dendrograms. This can be explained by the fact that situation dendrograms have more probabilities to be identical, since they contains less leaves, and the correlation coefficients (during the hierarchical clustering process) are computed on vectors of thousands components (the genes whose expression is measured in each situation). As a result, if we compare differently discretized gene expression matrix, the discretization thresholds for which we get the highest similarity score can be different for gene and situation dendrograms.

If we are interested in a unique similarity score, different solutions can be adopted. For example, we can consider the average between the gene and the situation similarity scores. A problem is that if one of the trees is totally dissimilar from the reference (relative score is equal to zero), the average value will not be zero. We can solve this problem by simply considering the square root of the product between the two similarity scores:

$$\overline{S_{AT}}(T, T_{ref}) = \sqrt{\overline{S_{GT}}(T, T_{ref}) \cdot \overline{S_{ST}}(T, T_{ref})} \tag{4}$$

where $\overline{S_{GT}}$ and $\overline{S_{ST}}$ and denote respectively the normalized similarity score for genes and situation, and $\overline{S_{AT}}$ denotes the average similarity score.

Following this definition, $\overline{S_{AT}}$ is always between the gene and the situation similarity score values. Furthermore, when at least one of the two similarity scores is equal to zero, also the average similarity score is zero.

3   Using Similarity Scores

Many discretization techniques can be used to encode gene expression properties from expression values that are either integer values (case for SAGE data [2]) or real values (case for microarray data [1]). In this paper, we consider for our experimental study only three techniques that have been used for encoding the over-expression of genes in [9]:

- "Mid-Ranged". The highest and lowest expression values are identified for each gene and the mid-range value is defined. For a given gene, all expression values that are strictly above the mid-range value give rise to value 1, 0 otherwise.
- "Max - X% Max". The cut off is fixed w.r.t. the maximal expression value observed for each gene. From this value, we remove a percentage X of this value. All expression values that are greater than the $(100 - X)\%$ of the Max value give rise to value 1, 0 otherwise.
- "X% Max". For each gene, we consider the situations in which its level of expression is in X% of the highest values. These genes are assigned to value 1, 0 otherwise.

We want to evaluate the relevancy of a discretization algorithm and its parameters according to the preserved properties w.r.t. a hierarchical clustering of the raw data. So, we have to compare the dendrograms obtained from the three different boolean matrices with the reference dendrogram.

We have considered three gene expression data sets: two microarray data sets and a SAGE data set. The first data set (CAMDA [24]) concerns the transcriptome of the intraerythrocytic developmental cycle of the plasmodium falciparum, a parasite that is responsible for a very frequent form of malaria. We have the expression values for 3 719 genes in 46 different time points, i.e., biological situations. The second data set (Drosophila [25]) concerns the gene expression of drosophila melanogaster during its life cycle. We have the expression values for 3 030 genes and 81 biological situations. The third one (human SAGE data from NCBI, see also [26,13]) contains the expression values for 5 327 human genes in 90 different cancerous and not cancerous cellular samples belonging to different human organs.

One indicator of the differences between derived boolean contexts is their density, i.e., the number of true values divided by the total number of cells in the matrices. In Fig. 3, we provide the density curves for the three data sets and depending on different thresholds for the "Max - X% Max" method. Notice that densities for the "X% Max" method are equal to X.

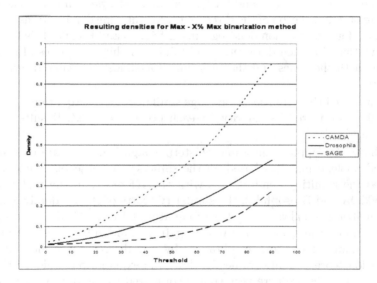

Fig. 3. Density values for different "Max - X% Max" thresholds

We processed all the computed boolean matrices with a hierarchical clustering algorithm based on the centered Pearson's correlation coefficient and the average linkage method. The same algorithm with the same options has been applied to the three original matrices. Finally, for each data set, we have compared all the genes and situations trees derived from the boolean matrices with the reference trees. The results in terms of **TScore** (Equation 1) for the "Mid-Ranged" method, are summarized in Fig. 4.

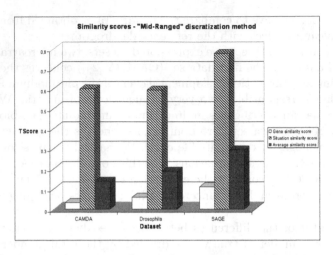

Fig. 4. Similarity scores for clustering trees on Mid-Ranged discretized matrices

For the "Max - X% Max" and "X% Max" methods, we summarize the results depending on the variation of the threshold X for the gene dendrograms in Fig. 5a and Fig. 6c, for the situation dendrograms in Fig. 5b and Fig. 6d. It is important to observe that, for each data set, we obtained the highest values of similarity scores for both the genes and the situations for almost the same discretization thresholds.

We have used the definition of average similarity score (Equation 4), to identify a unique measure of similarity for each boolean context. Results are summarized in Fig. 7.

We have also applied the same clustering algorithm on various randomly generated boolean matrices based on the same sets of objects. Then, we have compared the resulting dendrograms with the reference. In the first two data sets (CAMDA and Drosophila), the similarity scores of the randomly generated boolean matrices are always very low or equal to 0. In the SAGE data set, given a density value, the gene scores resulting from randomly generated matrices are always lower than the ones obtained by any discretization method (while the situation scores are always negligible). One explanation could be that the discretized matrices are here very sparse compared to the ones we derive from the first two data sets (see Fig. 3). Using a low threshold to discretize such a matrix does not make sense: obtained scores are similar to the scores which are computed on random boolean matrices. Moreover, using a high threshold value X for the "X% Max" discretization method leads to similarity scores that are close to those obtained for randomly generated matrices, though still higher. We can observe the behavior of this particular SAGE data set in Fig. 8.

As we can see, each discretization method has a set of threshold values for which it produce relatively high results in terms of similarity scores. Obviously, depending on the analysis task, one method can be more adapted than the

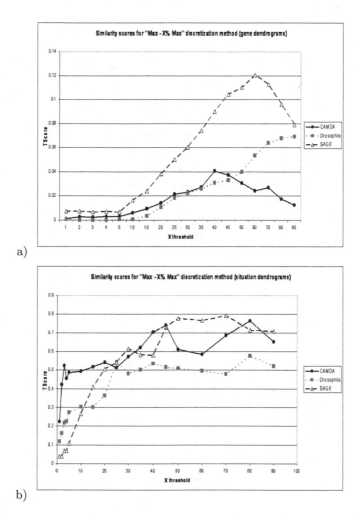

a)

b)

Fig. 5. Similarity scores w.r.t. different thresholds for "Max - X%Max"

other ones. For instance, even if both the "Max - X% Max" and the "X% Max" methods encode over-expression, the first one produces a boolean context whose density is strictly dependent on the maximal expression value for each gene. Instead, with the second method, we are sure that the density of the resulting boolean context is near to the X threshold. Does it mean that we are able to extract different kinds of patterns?

Clearly, the collections of patterns we can extract when using two different discretization techniques for over-expression encoding, will be different. We consider however that if we extract in proximity of the thresholds which produced the highest similarity scores for both methods, the intersection between the extracted collections will have a significant size. Patterns belonging to this intersection will also inform about rather strong associations.

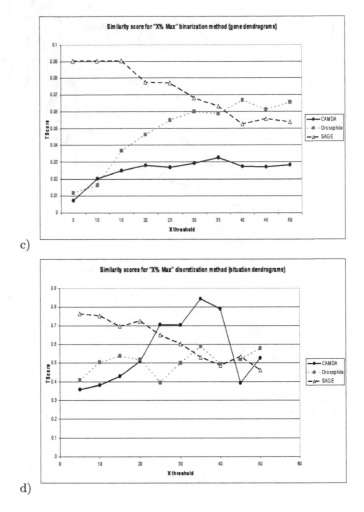

c)

d)

Fig. 6. Similarity scores w.r.t. different thresholds for "X%Max"

We have analyzed such intersections between different collections of formal concepts ([11]) which have been extracted from the boolean SAGE data set.

Definition 1. $(G, T) \in \mathcal{P} \times \mathcal{S}$ is a formal concept in $\mathbf{r} \subseteq \mathcal{P} \times \mathcal{S}$ when $T = \psi(G, \mathbf{r})$ and $G = \phi(T, \mathbf{r})$. ψ and ϕ are the classical Galois operators, i.e., we have $\phi(T, \mathbf{r}) = \{g \in \mathcal{P} \mid \forall o \in S, (g, o) \in \mathbf{r}\}$ and $\psi(G, \mathbf{r}) = \{o \in \mathcal{S} \mid \forall g \in G, (g, o) \in \mathbf{r}\}$. (ϕ, ψ) is the so-called Galois connection between $\mathcal{S}$ and $\mathcal{P}$. Notice that, by construction, when (G, T) is a formal concept, G and T are closed sets.

We used the D-MINER algorithm [14] to extract formal concepts under constraints: to avoid problems with outliers, we have considered formal concepts with at least 2 biological situations and at least 10 genes (i.e., $|G| \geq 10$ and

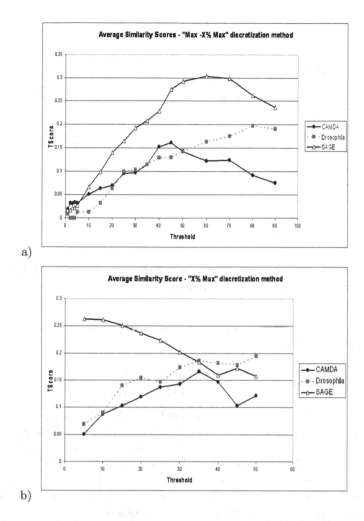

Fig. 7. Average similarity scores w.r.t. different thresholds for "Max - X%Max" (a) and "X%Max" (b)

$|T| \geq 2$). The mined boolean contexts have been obtained by the "Max - X% Max" and the "X% Max" over-expression encoding methods. We used the X threshold values which have produced the highest similarity scores (see Fig. 7). Then we compared all the collections extracted from each boolean context obtained with the first method, with all the collections related to the second method.

To compare pattern collections, we adapted the interactive self-similarity metrics introduced in [17]. Such a measure has been studied for comparing two collections of frequent itemsets extracted from two samples of a same data set. We modified it to work on formal concepts extracted from different boolean instances of a same data set.

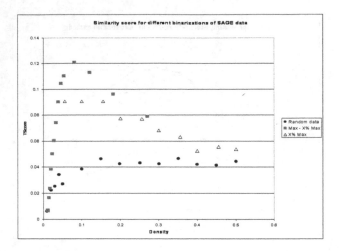

Fig. 8. Similarity scores w.r.t. density for "Max - X%Max", "X%Max" and random discretization methods on SAGE data

Given two boolean contexts $\mathbf{r}_1$ and $\mathbf{r}_2$ our pattern collection similarity measure is defined as follows:

$$Sim\,(\mathcal{C}_1, \mathcal{C}_2) = \frac{\sum_{x \in \{T_1\} \cap \{T_2\}} \frac{|\phi(x, \mathbf{r}_1) \cap \phi(x, \mathbf{r}_2)|}{|\phi(x, \mathbf{r}_1) \cup \phi(x, \mathbf{r}_2)|}}{|\{T_1\} \cup \{T_2\}|} \qquad (5)$$

where $\mathcal{C}_1 = \{(G_1, T_1) \mid (G_1, T_1) \text{ is a concept}\}$ and $\mathcal{C}_2 = \{(G_2, T_2) \mid (G_2, T_2) \text{ is a concept}\}$ are the collection of concepts extracted respectively from $\mathbf{r}_1$ and $\mathbf{r}_2$.

To better understand the meaning of this measure, we can see a toy example based on the tables Tab. 1b and Tab. 1b. Let $\mathcal{C}_b$ and $\mathcal{C}_c$ denote the collection of formal concepts extracted respectively from the boolean matrices in Tab. 1b and Tab. 1c (with a non empty set of genes and a non empty set of situations). The list of concepts contained in the two collections is:

$\mathcal{C}_b$	$\mathcal{C}_c$
$(G_{b1}, T_{b1}) = \{a, c, e\}, \{2, 4, 5\}$	$(G_{c1}, T_{c1}) = \{c\}, \{4, 5\}$
$(G_{b2}, T_{b2}) = \{b, f, g\}, \{1, 3, 5\}$	$(G_{c2}, T_{c2}) = \{b\}, \{1, 3\}$
$(G_{b3}, T_{b3}) = \{b, d, f, g\}, \{1, 3\}$	$(G_{c3}, T_{c3}) = \{b, d, f, g\}, \{1\}$
$(G_{b4}, T_{b4}) = \{a, c, e, h\}, \{4, 5\}$	$(G_{c4}, T_{c4}) = \{a, c, e, h\}, \{4\}$
$(G_{b5}, T_{b5}) = \{a, b, c, e, f, g, h\}, \{5\}$	

Clearly, only two sets of situations are shared by the two collections. They are $T_{b3} = T_{c2} = \{1, 3\}$ and $T_{b4} = T_{c1} = \{4, 5\}$. We get the following result:

$$Sim\,(\mathcal{C}_b, \mathcal{C}_c) = \frac{\frac{|G_{b3} \cap G_{c2}|}{|G_{b3} \cup G_{c2}|} + \frac{|G_{b4} \cap G_{c1}|}{|G_{b4} \cup G_{c1}|}}{7} = \frac{\frac{1}{4} + \frac{1}{4}}{7} = 0.07$$

Applying such a measure to our different collections gives the results collected in Tab. 2.

X %Max	Max -X%Max					
	40	45	50	55	60	65
2	0.009456	0.004353	0.001392	0.000412	0.000095	0.000016
5	0.147644	0.082908	0.028939	0.008899	0.002057	0.000334
8	0.093602	0.149451	0.146705	0.062045	0.017565	0.003033
10	0.033129	0.0663	0.131817	0.10268	0.039822	0.007915
15	0.003442	0.008034	0.026383	0.06342	0.097521	0.03868
20	0.000337	0.000792	0.0028	0.009689	0.035462	0.082248

Table 2. Self-similarity measures on different collections of concepts in SAGE data

Interestingly, the self-similarity values are relatively high in the intersection between the X values for which the "X% Max" method takes the highest similarity scores (**TScore**), and the X values for which the "Max -X% Max" method has the same behavior (see Fig. 7). We notice how the measures are usually very low (the highest one is about 0.15). It emphasizes the impact of the choice of a relevant discretization method. The relevancy of the extracted patterns is not only related to the preservation of some properties of the raw data set, but also tightly related to the specific biological problem at hand.

Comparing dendrograms resulting from the clustering of different types of derived boolean matrices enables to choose the "best" discretization method and parameters for a given data set. When looking at the average similarity scores for "Max - X% Max" and "X% Max" methods (see Fig. 7), we observe either an optimal value or an asymptotic behavior. It could mean that the best choice for the discretization threshold is a trade-off between the value for which we get the best similarity score and the value for which the data mining tasks remain tractable.

4   Robustness of the Measure

In Section 2, we proposed a method to assess gene expression property encoding. We refined the measure presented in [16] by defining an average similarity score which can take into account both gene and situation similarity scores. We now discuss the choice of the reference tree, and thus the choice of the clustering algorithm. Our idea is simple. If we apply a clustering algorithm with different parameters to the same gene expression matrix, and then compare all the resulting dendrograms using our method, the measures should be quite similar.

Even if there are methods that produce very similar results, and others that produce totally different results, the overall behavior of the measures should be identical, i.e., for each particular configuration of the clustering algorithm, the mean of the similarity scores obtained by comparing its resulting dendrogram with the dendrograms related to all the other configuration, should be high and should not differ too much from the means computed in the same way for the other configurations.

To perform the experiments, we have used the three datasets described in Section 2.3. Hierarchical clustering has been performed with the free software HCE 2.0 (Hierarchical Clustering Explorer) available on-line on the site of the Human-Computer Interaction Laboratory (University of Maryland)[1]. The used clustering metrics have been the classical Euclidean distance and the centered/uncentered Pearson's coefficients ([4]). Moreover, we used the four classical linkage methods (i.e., single, complete, average, average group linkage) and Shneiderman's 1-by-1 linkage method as well. For each data set, once the clustering process was completed, we have compared each of the resulting dendrograms with all the other dendrograms. This has been done for both gene and situation dendrograms. Due to space limitations, we provide only the average similarity scores for the Pearson's uncentered coefficient in the SAGE data set (see Tab. 3).

Metrics	Linkage	Average Similarity Scores - Pearson's Uncentered				
		Average	*Avg. Group*	*Complete*	*Single*	*Shneid.*
Pearson's	*Average*	1	0.67314383	0.67284944	0.80330149	0.73423766
Uncentered	*Average Group*	0.52915848	1	0.46868204	0.74420618	0.57334442
	Complete	0.72280557	0.64047742	1	0.76910562	0.65782797
	Single	0.37379950	0.44053048	0.33315260	1	0.38401040
	Shneiderman	0.69095298	0.68635693	0.57626332	0.77659575	1
Pearson's	*Average*	0.73765387	0.63184005	0.57791403	0.76935144	0.63659514
Centered	*Average Group*	0.51583859	0.71440599	0.71471718	0.73727445	0.58849284
	Complete	0.63213575	0.60668335	0.71471718	0.73727445	0.58849284
	Single	0.34417977	0.40501553	0.30670888	0.84271541	0.35339934
	Shneiderman	0.60198327	0.64004493	0.51441451	0.75143098	0.7112918
Euclidean	*Average*	0.22302538	0.26032110	0.21204947	0.34910165	0.22825697
	Average Group	0.22822833	0.26402535	0.20887047	0.34531201	0.23794469
	Complete	0.30246296	0.33102761	0.29277610	0.39226471	0.29859425
	Single	0.15444260	0.18272967	0.14310903	0.28929635	0.15970716
	Shneiderman	0.02970444	0.03884745	0.02795342	0.07367044	0.03641223

Table 3. Average similarity scores for clustering using Pearson's uncentered coefficient

Obviously, obtained values can be quite different. As expected, comparisons between "Pearson's coefficient" and "Euclidean distance" lead to rather low similarity scores. It is interesting to notice that comparisons with the single linkage method as reference leads to very high similarity scores. The same linkage method, compared with other references, give rise to rather low similarity scores. Our measure is indeed asymmetric.

We can compute the mean of the similarity scores obtained for each reference (see Fig. 9). The scores are always higher than the computed scores we got when comparing the dendrograms from the boolean matrices (see Section 2.3).

---

[1] http://www.cs.umd.edu/hcil/hce/

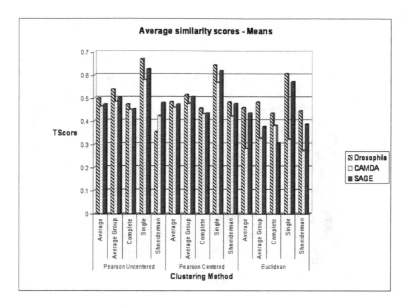

Fig. 9. Computed means of the average similarity scores for the three datasets

Finally, we have also considered the robustness of our metrics by looking at the overall behavior. For each data set, we have computed the mean of the measures shown in Fig. 9. To explain the content of this figure, let us remind all the steps of our analysis. First we have computed the similarity scores between all couples of computed dendrograms. Let T_i denotes the dendrogram resulting of a particular combination of clustering parameters ($i = 1..15$). Let S_{ij} denotes the similarity score computed between each couple of dendrograms T_i and T_j (T_i being the reference). Notice that in general $S_{ij} \neq S_{ji}$. In Fig. 9 we have the following values:

$$\overline{S}_i = \frac{\sum_{j=1}^{15} S_{ij}}{15}.$$

Let $\overline{S}_i^p$ denote the mean computed only on the dendrograms obtained by using the two Pearson's coefficient, i.e.,

$$\overline{S}_i^p = \frac{\sum_{j=1}^{10} S_{ij}}{10}.$$

For each data set, we are interested in the following measures:

$$\overline{S} = \frac{\sum_{i=1}^{15} \overline{S}_i}{15} \text{ and } \overline{S}^p = \frac{\sum_{i=1}^{10} \overline{S}_i^p}{10}.$$

Finally, we need to compute the standard deviations of the $\overline{S}_i$ and $\overline{S}_i^p$ values:

$$\sigma = \sqrt{\frac{\sum_{i=1}^{15} \left(\overline{S}_i - \overline{S}\right)^2}{15}} \text{ and } \sigma^p = \sqrt{\frac{\sum_{i=1}^{10} \left(\overline{S}_i^p - \overline{S}^p\right)^2}{10}}$$

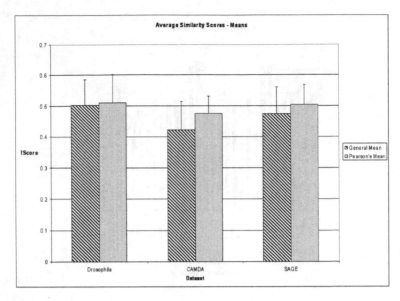

Fig. 10. Values of $\overline{S}$ and $\overline{S}^p$ and related standard deviations σ and σ^p

The final results are summarized in Fig. 10. Notice that the values of the means ($\overline{S}$ and $\overline{S}^p$) are near to 0.5 for every dataset, while the standard deviation is generally small. Both these observation make us conclude that the dendrogram resulting from the hierarchical clustering algorithm is a valid reference for our problem of comparing different method of gene expression property encoding. Moreover, the choice of the Pearson's correlation coefficient for the execution of the comparison (see Section 2.3), is shown to be adequate by the fact that the means computed only on the dendrograms obtained through this metrics ($\overline{S}^p$) are greater than the general means ($\overline{S}$), while the related standard deviations (σ^p) are similar or smaller than the general ones (σ).

5   Conclusion

We defined a new pre-processing technique that supports the evaluation and assessment of different discretization techniques for a given gene expression data set. The evaluation is based on the comparison of dendrograms obtained by clustering various derived boolean matrices with the one obtained on the raw matrix. The defined metrics is simple and we have validated its relevancy on different real data sets. A validation on a biological problem has been considered in [16]. This is a step towards a better understanding of a crucial pre-processing step when we want to apply the very efficient techniques based on set pattern mining from boolean data. Thanks to the exhaustive search for every pattern which satisfy the user-defined constraints, set pattern mining techniques like constraint-based mining of formal concepts appear to be complementary approaches to global pattern heuristic mining techniques like clustering.

Acknowledgements. The authors want to thank Céline Robardet, Sylvain Blachon and Olivier Gandrillon for the pre-processing of the SAGE data set, and Sophie Rome for her participation to microarray data preparation. Furthermore, we thank Claire Leschi and Jérémy Besson for their contribution to a preliminary version of this paper. Finally, this research is partially funded by ACI MD 46 (CNRS STIC 2004-2007) BINGO (Bases de Données Inductives pour la Génomique).

References

1. DeRisi, J., Iyer, V., Brown, P.: Exploring the metabolic and genetic control of gene expression on a genomic scale. Science **278** (1997) 680–686
2. Velculescu, V., Zhang, L., Vogelstein, B., Kinzler, K.: Serial analysis of gene expression. Science **270** (1995) 484–487
3. Piatetsky-Shapiro, G., Tamayo, P., eds.: Special issue on microrray data mining. SIGKDD Explorations, Volume 5, Issue 2 (2003)
4. Eisen, M., Spellman, P., Brown, P., Botstein, D.: Cluster analysis and display of genome-wide expression patterns. Proc. Natl. Acad. Sci. USA **95** (1998) 14863–14868
5. Niehrs, C., Pollet, N.: Synexpression groups in eukaryotes. Nature **402** (1999) 483–487
6. Boulicaut, J.F., Bykowski, A.: Frequent closures as a concise representation for binary data mining. In: Proceedings PAKDD'00. Volume 1805 of LNAI., Kyoto, JP, Springer-Verlag (2000) 62–73
7. Pei, J., Han, J., Mao, R.: CLOSET an efficient algorithm for mining frequent closed itemsets. In: Proceedings ACM SIGMOD Workshop DMKD'00, Dallas, USA (2000) 21–30
8. Zaki, M.J., Hsiao, C.J.: CHARM: An efficient algorithm for closed itemset mining. In: Proccedings SIAM DM'02, Arlington, USA (2002)
9. Becquet, C., Blachon, S., Jeudy, B., Boulicaut, J.F., Gandrillon, O.: Strong-association-rule mining for large-scale gene-expression data analysis: a case study on human sage data. Genome Biology **12** (2002)
10. Creighton, C., Hanash, S.: Mining gene expression databases for association rules. Bioinformatics **19** (2003) 79 – 86
11. Wille, R.: Restructuring lattice theory: an approach based on hierarchies of concepts. In Rival, I., ed.: Ordered sets. Reidel (1982) 445–470
12. Rioult, F., Boulicaut, J.F., Crémilleux, B., Besson, J.: Using transposition for pattern discovery from microarray data. In: Proceedings ACM SIGMOD Workshop DMKD'03, San Diego (USA) (2003) 73–79
13. Rioult, F., Robardet, C., Blachon, S., Crémilleux, B., Gandrillon, O., Boulicaut, J.F.: Mining concepts from large sage gene expression matrices. In: Proceedings KDID'03 co-located with ECML-PKDD 2003, Catvat-Dubrovnik (Croatia) (2003) 107–118
14. Besson, J., Robardet, C., Boulicaut, J.F.: Constraint-based mining of formal concepts in transactional data. In: Proceedings PAKDD'04. Volume 3056 of LNAI., Sydney (Australia), Springer-Verlag (2004) 615–624
15. Besson, J., Robardet, C., Boulicaut, J.F., Rome, S.: Constraint-based concept mining and its application to microarray data analysis. Intelligent Data Analysis journal **9** (2004) To appear.

16. Pensa, R.G., Leschi, C., Besson, J., Boulicaut, J.F.: Assessment of discretization techniques for relevant pattern discovery from gene expression data. In: Proceedings ACM BIOKDD'04 co-located with SIGKDD'04, Seattle, USA (2004) 24–30
17. Parthasarathy, S.: Efficient progressive sampling for association rules. In: Proceedings IEEE ICDM'02, Maebashi City, Japan (2002) 354–361
18. Moore, G.W., Goodman, M., Barnabas, J.: An iterative approach from the standpoint of the additive hypothesis to the dendrogram problem posed by molecular data sets. Journal of Theoretical Biology **38** (1973) 423–457
19. Robinsons, D.F.: Comparison of labeled trees with valency three. Journal of Combinatorial Theory, Series B **11** (1971) 105–119
20. DasGupta, B., He, X., Jiang, T., Li, M., Tromp, J., Zhang, L.: On distances between phylogenetic trees. In: Proceedings ACM-SIAM SODA'97. Volume 55. (1997) 427–436
21. DasGupta, B., He, X., Jiang, T., Li, M., Tromp, J., Zhang, L.: On computing the nearest neighbor interchange distance. In: Discrete mathematical problems with medical applications (New Brunswick, NJ, 1999), Providence, RI, Amer. Math. Soc. (2000) 125–143
22. Finden, C., Gordon, A.: Obtaining common pruned trees. Journal of Classification **2** (1985) 255–276
23. Cole, R., Hariharan, R.: An o(n log n) algorithm for the maximum agreement subtree problem for binary trees. In: Proceedings of the 7th annual ACM-SIAM symposium on Discrete algorithms, Atlanta, Georgia, United States (1996) 323–332
24. Bozdech, Z., Llinás, M., Pulliam, B.L., Wong, E., Zhu, J., DeRisi, J.: The transcriptome of the intraerythrocytic developmental cycle of plasmodium falciparum. PLoS Biology **1** (2003) 1–16
25. Arbeitman, M., Furlong, E., Imam, F., Johnson, E., Null, B., Baker, B., Krasnow, M., Scott, M., Davis, R., White, K.: Gene expression during the life cycle of drosophila melanogaster. Science **297** (2002) 2270–2275
26. Lash, A., Tolstoshev, C., Wagner, L., Schuler, G., Strausberg, R., Riggins, G., Altschul, S.: SAGEmap: A public gene expression resource. Genome Research **10** (2000) 1051–1060

Local Pattern Discovery in Array-CGH Data

Céline Rouveirol[1,2] and Francois Radvanyi[2]

[1] LRI , UMR 8623, Université Paris Sud, bât 490, 91405 Orsay cedex FRANCE
[2] Institut Curie, UMR 144, 26 rue d'Ulm 75248 Paris cedex 05 FRANCE

1 Local Patterns in Array-CGH Data

We report[3] in this paper about our practice of frequent pattern discovery algorithms in the context of mining biological data related to genomic alterations in cancer. A number of frequent item set methods have already been successfully applied to various biological data obtained from large scale analyses (see for instance [4] for SAGE data, [20,22,26] for gene expression data), and all of these have to face the peculiarity of such data wrt standard basket analysis data, namely that the number of observations is low wrt the number of attributes.

We handle in this paper a different type of biological data, complementary to gene expression data – and for the moment less studied – array-CGH[4] data [30]. Array-CGH allows measuring chromosomic alterations, i.e., gains and losses of segments of chromosomes in the genome of tumoral cells, compared with a normal control. It is hypothesised that a relatively low number of genomic alterations control the cancer process [15] and we expect that large scale mining of CGH array data will extract highly valuable information concerning genetic alterations, to be cross-checked with anatomo-clinical and gene expression data.

The measured phenomenon in CGH array is the number of copies of segments of chromosomes, that we will name *BACs* in the following, in a particular genome. In a "normal" situation, each segment occurs twice in the cell genome. In tumoral cells, a chromosome segment can be *amplified*, (i.e., it may be detected n times in the genome of the cell, with $n > 2$). Such sequences potentially contain *oncogenes*, i.e., genes responsible for tumour start and progression. Other zones are sometimes lost, in this case the genome of the tumoral cell may contain zero or one copy of that chromosome fragments. Such zones may contain candidate *tumour suppressor* genes, i.e., genes which inhibit tumour progression.

Moreover, and this will be our focus here, we expect to find combinations of such zones correlated with anatomo-clinical or biological attributes associated to the tumours (stage of tumour, probability of recurrence and progression, observed mutation of some genes, etc.). The identification of combination of regions of recurrent genomic alterations will have several implications in tumour biology. The definition of anatomopathological parameters like stage and grade of the tumours, is not accurate and is often "pathologist dependent". Also, the frontiers

---

[3] This work has been performed when the first author was on leave to CNRS in Institut Curie.
[4] CGH : *Comparative Genomic Hybridization.*

K. Morik et al. (Eds.): Local Pattern Detection, LNAI 3539, pp. 135–152, 2005.
© Springer-Verlag Berlin Heidelberg 2005

between two consecutive stages or grades are difficult to define. The definition of combination of altered regions associated with these anatomopathological parameters will give a molecular definition of these parameters which will be much less subject to controversy. The same is also valid for clinical parameters such as prediction of clinical outcome or response to therapy. The association of altered regions with anatomoclinical parameters or biological parameters (mutation of a gene or altered expression of a gene or a set of genes for example) is an important step towards the identification of genomic events necessary for tumour formation and progression. It will be of great help to identify signalling pathways activated in these processes. It is likely that the identified pathways will not be only involved in oncogenesis but also in other diseases, as well as in normal tissues. The combination of events that we will identified could be acquired in a stepwise fashion as originally thought. A much more recent theory suggests that combination of events are acquired simultaneously due to telomere dysfunction [12]. Our work will help to choose between these different mechanisms of tumour progression.

Looking for sets of gained or lost *BACs* correlated to a class attribute is a highly complex problem: even if array-CGH data is an order smaller than expression micro-array data, the number of *BAC* attributes is still very large when compared to the number of observations. We therefore do not address this problem directly, and proceed in two steps. We first look for *local constrained patterns*, i.e., patterns that both satisfy a set of relevant properties wrt genome biology and some nice computational properties. For instance, one such constraint states that only *sequences* (instead of itemsets) of *BACs* are relevant for the problem at hand. Another constraint sets that we are only interested in *closed sequences*, i.e., the largest sequences that occur in a given set of observations. These patterns are *local* because they are expected to characterise a relatively small number of observations. As no global model for such a complex problem as cancer initiation and progression is available, our way to proceed is rather to assemble jigsaw pieces these local patterns allows us to extract. These local patterns are then used as a *filter* to re-describe the initial data before learning discriminant patterns, similarly as what Kramer and De Raedt proposed in [18].

Such re-description or abstraction of the mining task in terms of the extracted constrained patterns allow us to enforce the language bias, which is far too weak in the initial learning problem to get any result in reasonable time. The hypothesis space after abstraction is much smaller, it is therefore possible to exhaustively explore it, with the goal of finding discriminant patterns wrt given class attributes. In other words, we see local patterns as a way to strongly bias the set of expected solutions (a *point de vue* on the data), so that exhaustive exploration of this viewpoint and its limits wrt the discrimination problem at hand is possible. We do not expect to build a classifier based on array-CGH data only with a high recall, as it is highly unprobable that genomic alterations are sufficient to fully explain, for instance, the stage of a tumour, as many other genetic events other than loss, gain or amplification [1] as well as epigenetic events take place during cancer initiation and progression. However, we are interested in isolating some subgroups of patients (even small ones), for which exists a pattern of gained and lost regions is significantly correlated to attributes such as the stage of tumour or specific gene mutations.

The remainder of the paper is organised as follows. Section 2 introduces the learning problem and motivates the reformulation in terms of local patterns, namely *GL regions*. Then, section 3 describes more formally the reformulation step and framework. Section 4 focus on the mining step from reformulated data. Finally, section 5 provides some experimental results and section 6 concludes and draws some perspectives for this line of research.

2   The Mining Tasks

2.1   The Initial Data and Mining Problem

Raw CGH-array data are numerical matrixes, giving for each *BAC* of a given array, an intensity ratio, which, provided that there are no measurement errors and that all cells for the test sample are all tumoral, should be equal to $\frac{n}{2}$, where n is the number of copies of the *BAC* chromosome segment in the tumoral DNA (see figure 1). In a normal situation, the cells all contain 2 copies of the *BAC*, a log2 ratio of 0 is observed for this *BAC*. A -0.5 log2 ratio means that a single copy of the *BAC* has been detected in the genome cells of the sample, a 0.5 log2 ratio means that a single copy of this *BAC* has been gained (the cell therefore contains 3 copies of it). High positive or negative deviations denote the gain or loss of more than one copy of a given *BAC*.

In this work, the array-CGH data has been normalised and discretised, using the algorithm *GLAD* (see [16] for a full description of the method). Let us just mention here that *GLAD* extracts a stepwise function from each CGH profile, and that this step function is then matched to a set of discrete values (see figure 1)[5]. It is more natural to discretise CGH data than gene expression data, as the underlying phenomenon in CGH data (i.e., the copy number of a set of *BACs* in tumoral cells) is discrete. Moreover, the discretisation steps makes it possible to apply a whole range of well studied discrete algorithms, which are potentially interesting alternatives to numerical techniques, such as statistical clustering or Support Vector Machines [2]. We will assume in the remainder of the paper that each *BAC* is coded as *Gained*, *Normal* or *Lost*. Our input data can be formally described as follows.

Definition 1. *Given a set of* BACs *$\mathcal{B}$, and a set of observations $\mathcal{O}$, an extraction context is a boolean matrix M of $2 * |\mathcal{B}| = N_B$ boolean attributes (two boolean attributes are necessary to encode the status of each* BAC*, gained, lost or normal) and $|\mathcal{O}| = N_O$ observations representing gained (resp. lost)* BACs *for each observation of $\mathcal{O}$. $\mathcal{O}$ can be partitioned into two groups $\mathcal{O}^+$ and $\mathcal{O}^-$, yielding M to split into M^+ and M^-, named in the following the positive and negative contexts.*

---

[5] Courtesy of Ph. Hupé et al.

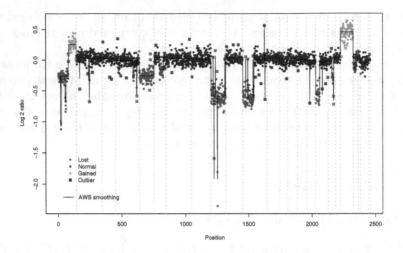

Fig. 1. Example of a whole genome CGH profile of a tumoral DNA. The X axis denotes the position of *BACs* on the genome, the Y axis denotes the log2-ratio of the estimated copy number for *BACs*.

From this input, we wish to build discriminant patterns that are, in a first approximation, sets of *BACs* which are both frequent in M_+ and infrequent in M_-. These discriminant patterns are genomic alteration signatures, character- istic for the M_+ context. More formally, the problem of finding discriminant sequence patterns for two contexts M_+ and M_- can be formulated as follows.

Definition 2. *Given a* BAC *attribute set* $\mathcal{B} = \{bg_1, \ldots, bg_{N_B}, bl_1, \ldots, bl_{N_B}\}$ *and two boolean contexts* M_+ *and* M_- *built on* $\mathcal{B}$, *we wish to find all sets of* BAC *attributes* dp_i *such that* $freq(dp_i, M_+) \geq min\_t$ *and* $freq(dp_i, M_-) \leq max\_t$.

For solving the above task, we have to cope with the problem, traditional when handling micro-array data, of handling contexts with a number of attributes much larger than the number of observations (most datasets available describes about a hundred observations in terms of about thousands of *BACs*). Several ap- proaches have been implemented in bioinformatics to overcome this dimensional- ity problem. Let us, among others, mention the Biorelief system [21] that selects most discriminant genes wrt to a given class. Another approach, introduced in [26], and developed in [5] makes use of the Galois lattice connection properties in order to efficiently search the power-set of observations, much smaller in the case of microarray data than the power-set of attributes which is traditionally searched by frequent itemset techniques. We follow here another strategy, that we describe in the next sections, for coping the dimensionality problem.

2.2   Local Patterns for Learning Task Abstraction

The strategy we adopt is similar of what has been performed in [18,13] in other biological application contexts: first identify *constrained local patterns* and then use them to reformulate the initial learning problem into an efficiently solvable problem. This local pattern reformulation step is indeed very problem dependent as its goal is to select or construct salient features for the redescribing the learning problem. We introduce in the following the definition of constrained pattern.

Most often in the Data Mining community, *local patterns* in an extraction context are *subsets* of attributes that occur frequently in this context. One specificity of this application is that $\mathcal{B}$ is totally ordered with $\prec_B$, the ordering of *BACs* on the genome (see fig. 1). Our *local patterns* in the following are *sequences* of *BACs*, and more specifically, *BACs* sequences that all have the same status of alteration (gained or lost) in a sufficient number of observations in our extraction context.

Definition 3. *A sequence of* BACs *is an interval of* BACs *given* $\prec_B$. *In the following, we denote by* $S(\mathcal{B})$ *the set of all sequences of* $\mathcal{B}$. *The support of a sequence s of* $S(\mathcal{B})$ *in a context* M, *denoted* $supp(s, M)$, *is the set of observations of* M *that contain* s *(i.e., such that all attributes of* s *are equal to 1 in those observations). The frequency of* s *in* M, *denoted* $freq(s, M)$, *is the size of its support.*

$S(\mathcal{B})$ is partially ordered by $\subseteq$, namely interval inclusion. $(S(\mathcal{B}), \subseteq)$ is a lattice, much smaller than the lattice of itemsets of $\mathcal{B}$, isomorphic to the powerset[6] of $\mathcal{B}$. Sequences of *BACs* recurrently altered, i.e. sequences of altered *BACs* that occur more than a minimal threshold in a sample of array CGH profiles, can be numerous especially if the minimum threshold is low and not all of them are necessarily relevant. It may be useful in that case to introduce a more complex notion of *constrained local pattern* that we detail in section 3.1.

Once such constrained local patterns $\mathcal{S} = \{s_1, \ldots, s_{N_S}\}$ have been identified, we can now combine them in order to build discriminant complex patterns, i.e., sets of constrained local patterns which are both frequent in a dataset M_+ and infrequent in a dataset M_-. These discriminant sequence patterns are genomic alteration signatures, characteristic for the M_+ context.

2.3   Reformulation of Input Data

Given a set of constrained local patterns $\mathcal{S} = \{s_1, \ldots, s_{N_S}\}$ that satisfy a conjunction of properties, we reformulate the initial contexts M_+ and M_- into much compact and simpler contexts M_{+r} and M_{-r} as follows. Both reformulated contexts M_{+r} and M_{-r} have as attributes constrained local patterns of $\mathcal{S}$. For each observation o of M (M_+ or M_-), and for each constrained local pattern $s_i = [lb_i..ub_i]$ of $\mathcal{S}$, if all[7] *BAC* attributes b_j of the constrained sequence of *BACs* ($lb_i \preceq_B b_j \preceq_B ub_i$) occur in o, then $M_r(s_i, o) = 1$.

The obtained extraction context M_r can be seen as an *abstraction* of the initial one M [28]. M_r is much simpler than M: firstly, it has substantially

---

[6] $|S(\mathcal{B}) = \frac{N_B * (N_B - 1)}{2}$ to be compared with 2^{N_B}.

[7] One might consider here a more subtle reformulation, i.e. $M_r(s_i, o) = 1$ if a fixed percentage of the *BAC* attributes of s_i occurs in o.

less attributes (see section 5 for more details) than M; secondly, the set of reformulated observations may also be reduced if some observations of M do not contain any of the GL regions of R. More formally, the problem of finding discriminant sequence patterns for two datasets M_+ and M_- can be formulated as follows.

Definition 4. *Given constrained GL region set* $S = \{s_1, \ldots, s_{N_S}\}$ *and two boolean contexts* M_{+r} *and* M_{-r} *described in terms of* S *attributes, we wish to find all sets* dp_i *of* S *such that* $freq(dp_i, M_+) \geq min\_t$ *and* $freq(dp_i, M_-) \leq max\_t$.

The reformulation step has numerous (and yet to be formally evaluated) consequences on the extraction of discriminant complex patterns. Obviously, searching for discriminant patterns in this new search space of sets of constrained sequences of *BACs* is much more efficient. We will see in section 5 that the number of constrained sequences attributes is now in the same order as the number of observations. There is of course a loss of information during this process: further complex pattern extraction steps can only build combinations of constrained gain and loss regions wrt a class attribute. We describe in the next section an instance of reformulation scheme.

3   Local Constrained Patterns

We first specify our definition of constrained local patterns implemented in this application and give an example of one such reformulation step.

3.1   Recurrent Loss and Gain Regions

Our local patterns, named *Gain/Loss regions* or *GL regions* for short in the following, are sequences of *BACs* that are recurrently gained or lost in our extraction context.

Definition 5. *Given a context* M, *describing a set of observations* O *in terms of a set of totally ordered attributes* $S(\mathcal{B})$, *a GL region* $r \in S(B)$ *is a set of contiguous BACs (i.e. a sequence of* BACs*), which are all gained or lost in at least one observation of* M. *Importantly, a region* r *is a* closed *sequence of* $S(\mathcal{B})$, *i.e., there is no super-sequence of* r *which has the same support as* r.

Efficient generation of closed itemsets has been studied a lot (see, among others [23,24]) because they are a concise representations for itemsets. This latter is true as well for sequences, although generation of closed sequences has been les studied in the Data Mining community. As this is the case for frequent patterns in dense contexts [3], there can still be a high number of frequent closed sequences, specially if the min frequency threshold is relatively low (around 5 or 10%). Moreover, we cannot only rely on minimal support pruning to ensure the relevance of the obtained regions. We therefore use a rich set of additional constraints on sequences of *BACs* in order to be able to construct potentially relevant low frequency regions. We have selected the following constraints:

- min or max bound on the percentage of observations in the support having a given property (such as being an outlier, being unmeasured, belonging to a given class, etc.)
- the region is n-right (left) bounded, i.e., if the region is right (left) delimited in n observations at least.

Let us mention here that closeness, in combination with other constraints should be handled with care [8], if *all* closed constrained sequences of $S(\mathcal{B})$ should be enumerated. Finally, if constrained regions are still too numerous, we might be only interested into *minimal constrained* regions. A constrained region r is minimal if there is no other constrained region r' such that $r' \subset r$.

Example 1. Let us assume that we have the following context M:

	b_1	b_2	b_3	b_4	b_5	b_6	b_7	b_8	b_9	b_{10}	b_{11}
o_1	1	0	0	0	1	1	0	1	1	1	1
o_2	1	1	0	0	1	1	0	0	0	1	0
o_3	0	0	1	0	1	1	1	0	1	0	0
o_4	1	0	0	0	1	1	0	1	1	1	0
o_5	1	1	0	1	1	1	0	1	1	1	0
o_6	0	0	0	0	1	1	1	0	1	0	1
o_7	0	0	0	0	1	1	0	1	0	0	0
o_8	0	0	1	1	1	1	0	1	1	1	0

The set of minimal regions with minimal frequency in the context of 3 and bounded left and right by at least 2 examples is $\{[b_1..b_1], [b_3..b_3], [b_5..b_6], [b_8..b_{10}]\}$.

We have designed [27] an algorithm that computes all minimal constrained GL regions, given a three valued context (Gain/Normal/Loss). We will not describe it here, but we will rather focus on assessing how such local patterns can make the extraction of discriminant rules from our genomic data tractable and efficient.

4   Mining the Reformulated Data for Discriminant Patterns

For this second step, our language of patterns is that of *itemsets of constrained GL regions*. This mining task can be seen as finding patterns of GL regions that satisfy the conjunction of a monotone and an anti-monotone constraint. As the monotone and anti-monotone constraints do not apply on the same context, it has not been possible to make the evaluation of both monotone and anti-monotone constraints more efficient by reducing the input context as done in Examiner [7].

Discriminant patterns as described in definition 4 are a special case *emergent* patterns given two data collections [14]. The growth rate of a pattern p, considering two contexts M_+ and M_- is the ratio $\frac{freq(p,M_+)}{freq(p,M_-)}$, and an emergent pattern is a pattern the growth-rate of which is higher than a fixed threshold. The algorithm described in [14] relies on the computation of maximal bounds both for the anti-monotone constraint and the negation of the monotone constraints. The computation of such maximal bounds was empirically evaluated as too computationally expensive, specially when mining with low min frequency thresholds, both in the positive and negative contexts. More recent emerging patterns algorithms [31] might be a solution to that problem. In the meantime, we have implemented and adapted the *Dualminer* algorithm [10], because it is flexible and it allows the simultaneous handling of monotone and anti-monotone constraints. We remind here the main algorithms of *Dualminer* and their adaptation to solve our learning task.

Dualminer explores a state graph, the nodes of which are sub-algebras of itemsets of S. Each sub-algebra *state* is a set of itemsets which are by delimited by a lower and upper bound. Each state *state* is represented by:

- $IN(state)$: the upper bound of *state*, it contains attributes that belong to *all* itemsets of *state*
- $OUT(state)$: attributes that *do not* belong to *any* itemset of *state*. The complement of $OUT(state)$ in S is the lower bound of *state*
- $CHILD(state)$: candidate attributes for expansion of *state*.

Notice that $IN(state) \cup OUT(state) \cup CHILD(state) = S$, i.e. the set of all GL region attributes. Let $compl(s)$, where $s \in S$, denote the complement set of s in S, i.e. $S \setminus s$.

In the following, P sets a minimum frequency for GL region patterns in the positive context M_{r+} (anti-monotone constraint) and Q dually sets a maximum frequency for GL region patterns in the negative context M_{r-} (monotone constraint). A leaf of the search space is a state such that $CHILD(state) = \emptyset$.

Example 2. The context of example 1, given the set S computed and the fact that observations o_1 to o_4 are labelled as positive and observations o_5 to o_8 are labelled as negative yields the following contexts M_{r+} and M_{r-} on $S = \{s_1, s_2, s_3, s_4\}$:

	s_1	s_2	s_3	s_4
o_1	1	0	1	1
o_2	1	0	1	0
o_3	1	1	1	1
o_4	0	0	0	1

	s_1	s_2	s_3	s_4
o_5	1	0	1	1
o_6	0	1	1	0
o_7	0	0	1	1
o_8	0	1	1	1

Given the constraints $P = freq(p, M_{r+}) \geq 2$ and $Q = freq(p, M_{r-}) \leq 1$, the solution of the mining task is the sub-algebra $< s_1, s_1 s_3 s_4 >$, that stands for the three GL region patterns $\{s_1, s_1 s_3, s_1 s_3 s_4\}$.

Search starts from a state with empty IN and OUT sets and with a $CHILD$ set containing all attributes of S. Each IN set of a state to be explored *state* is first checked against P. If $IN(state)$ does not satisfy P, search jumps to the next unexplored state. Otherwise, *state* is simplified by iterative calls to *Mono_prune* and *Anti_prune* algorithms (see figure 2) until no more pruning can be performed. When the state cannot be further simplified, it is checked against Q. If it satisfies Q, $IN(state), compl(OUT(state)$ is a solution, and search proceeds directly to the next unexplored state. Otherwise, *state* is expanded into two descendants: its left descendant is equal to *state* except for one element of $CHILD(state)$ that moves to its IN set, whereas the second descendant is obtained by moving the same element to its OUT set. Search stops when no more state can be developed.

Anti_prune prunes the $CHILD$ set of the state, by adding to OUT any attribute x such that $IN(state) \cup x$ does not satisfy the anti-monotone constraint P (i.e., $freq((IN(state) \cup x), M_{r+}) < min\_threshold$). *Mono_prune* transfers from the $CHILD$ set to the IN all attributes x such that $compl(OUT(state) \cup x)$ does not satisfy Q, i.e., $freq(compl(OUT(state) \cup x), M_-) > max\_threshold$. Both prunings are mutually dependent : an addition to the IN set triggers *Anti_prune*, whereas an addition to the OUT set triggers *Mono_prune*, until IN and OUT reach a fixed point.

```
Anti_prune (state)
  For all x ∈ CHILD(state) do
      If IN(state) ∪ x does not satisfy P then
    CHILD(state) := CHILD(state) - x
    OUT(state) := OUT(state) ∪ x

Mono_prune (state)
  For all x ∈ CHILD(state) do
      if compl(OUT(state) ∪  x) does not satisfy Q then
          CHILD(state) := CHILD(state) - x
          IN(state) := IN(state) ∪ x
```

Fig. 2. Monotone and anti monotone pruning

Example 3. After example 2, the search starts from the state $state_i$ such that $IN(state_i) = OUT(state_i) = \emptyset$, and $CHILD(state_i) = S = \{s_1, s_2, s_3, s_4\}$. *Anti_Prune* moves s_2 from $CHILD(state_i)$ to $OUT(state_i)$ and *Mono_Prune* moves s_1 to $IN(state_i)$. $state_i$ cannot be further simplified. $IN(state_i)$ satisfies both P and Q, therefore one solution sub-algebra is $< IN(state_i) = s_1, compl(OUT(state_i)) = s_1 s_3 s_4 >$. Search then stops because no more state can be generated.

4.1   Constructing Classifiers

A sub-algebra result of *Dualminer* concisely describes a set of patterns satisfying both constraints P and Q of the problem. In our case, assuming a boolean class attribute C, whose positive examples belong to M_+ and whose negative examples belong to M_-, *Dualminer* builds a set of discriminant GL region patterns for C. As noted in [19], each isolated discriminant pattern has a very poor accuracy and recall. Learning a classifier for C with low recall and using the default classification rule that if none of the patterns for C recognises a new observation, it is classified as not belonging to C, would lead to a very large prediction error. Therefore, we have adopted the strategy of building a classifier both for concept C and its complement that we denote $notC$, in order to bound the prediction error as much as possible (no classification is better than a systematic error). Given the positive and negative contexts (M_+, M_-) and the min and max frequency thresholds $(smin_+, smax_-, smin_-, smax_+)$, our algorithm builds:

- a set of sub-algebras representing all GL region patterns p satisfying both $freq(p, M_+) \geq smin_+$ and $freq(p, M_-) < smax_-$: these complex patterns are characteristic for observations of M_+;
- dually, a set of sub-algebras representing all GL regions patterns p' satisfying $freq(p', M_-) \geq smin_-$ and $freq(p', M_+) < smax_+$: these complex patterns are characteristic for M_-.

The upper bound (i.e., the IN sets) of each resulting sub-algebra is selected and tested for statistical significance, performing a χ^2 test on the 2x2 contingency table for the pattern wrt class C, with threshold of 3.84 (corresponding to a 95% p-value). A final selection step only keeps the most general significant patterns. We denote in the following $p_{i+}, 1 \leq i \leq N_{P+}$ as the set of patterns characteristic for class C and $p_{j-}, 1 \leq j \leq N_{P-}$ as the set of characteristic patterns for class $notC$.

A new observation (whose class is unknown) is classified as follows: all patterns for C and $notC$ are checked on the new observation. The observation gets a vote for class C each time there is a pattern p_{i+} for class C such that $p_{i+} \subseteq o$. Symmetrically, o gets a vote for $notC$ each time there is a pattern p_{j-} for class $notC$ such that $p_{j-} \subseteq o$. The class of o is¡ computed as the majority class for all votes. If there is no pattern for C or $notC$ that recognises (i.e., is included in) observation o, or if o gets the same number of votes for both C and $notC$, the observation is not classified.

5   Validation

We have applied the chain (extraction of constrained GL regions - reformulation - extraction of discriminant GL patterns) we have just described to a array-CGH dataset about bladder cancer patients. The dataset contains 92 CGH arrays (8 normal samples and 84 tumoral samples) described in terms of about 2400 BAC attributes, about equally distributed on 24 chromosomes (22 + chromosomes X and Y). We have selected two possible class attributes in this application, the *stage* of the tumour and whether a specific gene, namely *FGFR3*, is mutated in the tumour or not. The stage corresponds to the depth of invasion by the tumour cells of the surrounding tissue (T_a: papillary tumour with no invasion of the basal membrane, T_1: tumour which invades the basal membrane but not the underlying smooth muscle, T_2: tumour which invades the smooth muscle, T_3: tumour which invades perivesical tissue, T_4: tumour which invades a neighbouring organ). Increasing index of stage corresponds to an increasing severity of cancer. *FGFR3* mutation is already known to be an important feature for bladder cancer, as it has been frequently observed, most often in non invasive cancers) [6].

Stage of tumour is available for all tumours, and split into 26 T_a, 16 T_1, 10 T_2, 14 T_3 and 18 T_4, we will look here for patterns characterising invasive tumours (*stage* $= T_i, i \geq 2$, 42 of them) against non invasive tumours (*stage* $= T_a$ or T_1, 42 of them also) and vice-versa. *FGFR3* mutation is available for 73 tumours (29 mutated, 44 not mutated).

5.1   Computing Region Attributes – Reformulating Data

GL regions have been computed with different minimal frequency thresholds: 5% (4) and 10% (8) in the tumoral samples, different minimal number of observations defining the regions. Finally, we selected minimal GL regions only, or all GL regions satisfying the above constraints. Although selecting the best parameters for this step (and thus, the best abstraction step for mining discriminant genomic patterns) is crucial, we will not address it here. We will rather present some figures for the best results obtained by leave-one-out validation, both for the *FGFR3 mutated/non mutated* and the *invasive/non invasive tumour* classification tasks, obtained with the following parameter values: *5% min frequency threshold, 2 observations minimum for delimiting a region and minimality of regions*. In this configuration, about 70 GL regions are obtained.

When increasing the min frequency threshold to 10%, only about 35 minimal GL regions are computed. A similar behaviour is observed when strengthening the constraint on the minimal number of observations "defining" a region. On the other hand, if the minimality constraint on GL regions is relaxed, 90 GL regions are computed. We give some yet intuitive reasons why all these configurations yield to worst prediction results. Finally, note that, in the best configuration described above, 13 observations (over 84) are lost because none of the GL regions computed occur in those observations.

5.2   Computing Discriminant Patterns

For evaluating the performance of combining GL region reformulation and discriminant pattern learning, we have performed a leave-one-out validation. Minimal GL regions are extracted from a training set of size $N_O - 1$, the training set is then reformulated given the GL regions, as well as the test example. The test example is then classified given the discriminant GL region patterns extracted from the reformulated training set. This process is repeated for all observations in the context. We provide in the following some figures in the tables in the annex as well as in the figures 3, 4 and 5. The discriminant patterns provided in figures 6 are the most stable ones, i.e. those that were most frequently obtained over the different runs of the leave-one-out validation.

Figure 3 shows that the predictions results are relatively good for the task of discriminating *FGFR3* mutated samples against non *FGFR3* mutated samples. For all values of $smin_+$ and $smin_-$, we observe that learning patterns without

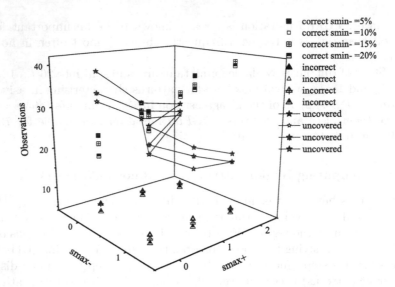

Fig. 3. Leave-one-out evaluation of discriminant patterns for *FGFR3* mutation, with $smin_+ = 5\%(2)$. $smin_-$ varies from 5% (3) to 20% (9), $smax_+$ and $smax_-$ vary between 0 and 2. The number of negative (resp. positive) examples is not allowed to be greater than half the number of positive (resp. negative) examples.

exceptions ($smax_+ = smax_- = 0$) yields bad predictive results. This supports the fact that data are noisy. In particular, increasing $smax_-$ from 0 to 1 increase the number of correct predictions by $1/3$. Best performances are obtained with $smax_- = 1$ and $smax_+ = 2$. In both configurations of figures 3 and 4, the number of incorrectly classified examples is very low wrt the number of well classified examples and a majority of those examples are false negatives. In those two configurations, the number of unclassified examples is not negligible, but still significantly lower than the number of well classified examples. It follows that our classifier based on patterns of gains and loss of chromosomes fragments is a reasonable predictor for *FGFR3* mutation status. The best prediction (43 well classified examples) is reached for $smin_+ = 5\%$, with $smin_- = 15\%$, $smax_- = 1$ and $smax_+ = 2$ and for $smin_+ = 20\%$, with $smin_- = 10$ or 15%, $smax_- = 1$ and $smax_+ = 2$. All other parameters being fixed, increasing $smin_+$ does not significantly improve the prediction performance (see figures 3 and 4). This means that there are few predictive patterns associated to *FGFR3* (see figure 6) and that these patterns occur with a high frequency in M_+.

The landscape is quite different when attempting to find discriminant GL regions patterns for class *invasive* vs *non invasive*. First, the number of unclassified examples is much greater than for the previous discrimination problem (in the best configuration, similar to the number of well classified examples).

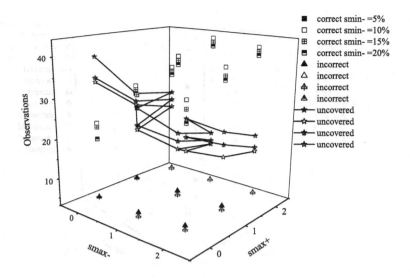

Fig. 4. Leave-one-out evaluation of discriminant patterns for *FGFR3 mutation* with $smin_+ = 20\%(8)$. $smin_-$, $smax_+$ and $smax_-$ vary as in figure 3

The reason for this behaviour is that there are very few GL region patterns for non invasive tumours, and all with low support. If $smin_+ \geq 10\%$, the set of patterns for the class of *non invasive* tumours is empty, the prediction performances are those of patterns for invasive tumours only. The prediction performance is insensitive to value of $smax_-$, but improves significantly when $smax_+$ increases from 0 to 1, and then degrades when increasing $smax_+$ to 2 (the number of false negative increases), all other parameters being fixed. No prediction is correct if $smin_-$ is equal to 20%. The best performance, although more modest as that obtained for *FGFR3* mutated vs non mutated, is obtained for $smin_+ = smin_- = 10\%$ and $smax_+ = 1$. Let us notice that some of the patterns obtained, such as the first rule for invasive tumours (see figure 6) have already been reported in the literature [11]. Further evaluation of the rules is currently going on.

Finally, we have tested the whole (reformulation + discriminant learning) process on different reformulations of the initial data. On one hand, the minimality constraint for GL regions has been relaxed, on the other hand, we have computed minimal GL region attributes with a min support threshold of 10% or the number of observations delimiting a region has been increased. In all cases, consistently worst prediction results were obtained. In the first case, discriminant patterns had a slightly worse prediction. This is certainly due to the fact that some of the GL region patterns selected after the χ^2 test were built using some of the more specific (non minimal) GL regions : as a result, the resulting classifiers loose some of their prediction. In the latter case, the prediction performance degraded significantly, in particular due to the fact that the number of

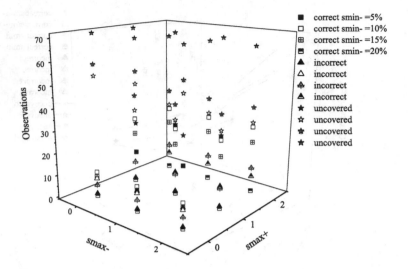

Fig. 5. Leave-one-out evaluation of discriminant patterns for stage, with $smin_+ = 10\%(4)$

false negatives increased at lot. This certainly indicates that a too stringent selection on GL regions has taken place and that some information relevant to the
discrimination process has been lost. These results fully demonstrate, without
surprise, that mining steps after reformulation are very sensitive to the quality
of extracted local patterns. This is why the language to describe the constraints
on those local patterns has to be as rich as possible.

Discriminant patterns we obtained for classes invasive vs. non invasive and
FGFR3 mutated vs. non mutated are less than a hundred, and this small number
makes it possible to think about delivering it to an expert for inspection.

6   Conclusion

We have described in this paper a method that learns discriminant patterns from
discrete data representing the number of occurrences of a sets of chromosome
segments (i.e., *BACs*) in DNA extracted from different tumours of the same
type. This method decomposes the resolution of extraction problem in which the
number of attributes is high wrt to the number of observations into two steps :
first, finding a relevant reformulation (here, reccurrent Gain/Loss sequences on
the human genome) that allows abstracting the initial problem into a simpler
one, that still retains important features from the initial problem but has a much
smaller number of attributes. There is of course a high number of such possible
reformulations, and we have relied on biological knowledge to select those we
have evaluated in this paper.

```
if RP11-28N6 - RP11-118C13 [9p24.3] is lost and
   RP11-14J9 - CTD-3145B15 [9q21.13] is lost then FGFR3 is mutated
if RP11-18B9 - RP11-18D13 [11p12] is lost and
   RP11-14J9 - CTD-3145B15 [9q21.13] is lost then FGFR3 is mutated
if RP11-102K7 - RP11-10G10 [8q22.2] is gained then FGFR3 is not mutated
if RP11-40D5 - RP11-40D5 [4q28.3] is lost then FGFR3 is not mutated

if CTD-2202A14 - CTD-2202A14 [5q32] is lost and
   RP11-28N6 - RP11-118C13 [9p24.3] is lost then tumour is not invasive
if RP11-32H11 - RP11-32H11 [10q26.2] is lost then tumour is invasive
if RP1-32G10 - RP1-32G10 [Yp11.31] is lost and
   RP1-24H17 - RP1-24H17 [5p15.33] is gained then tumour is  invasive
if RP11-240A17 - RP11-50E20 [8p23.3] is lost and
   RP11-102K7 - RP11-10G10 [8q22.2] is gained then tumour is invasive
```

Fig. 6. Excerpt of discriminant patterns for *FGFR3 mutated* vs. *non mutated* tumours and for *non invasive (stage $< T_2$)* vs. *invasive* tumours *(stage $\geq T_2$)* in bladder cancer. GL regions are denoted by their starting and ending *BACs*. The location of the *BAC* sequence on chromosomes is given into brackets.

This approach of looking for *all* discriminant patterns in a given langage is quite opposed to the strategy of many heuristic machine learning (such as decision tree systems [25]) that build a single set of rules, from a tree built by selecting at each node the test that optimizes a heuristic measure, such as information gain. The model that we have of our application domain is too weak, and, in particular, we have too few hints about the underlying distribution of our observations, to use such methods. We therefore rely on exhaustive symbolic methods and on experts of the domain that can interpret and evaluate the results. The type of approaches we have presented here can be computationally expensive [9], but it provides results which are relatively easy to interpret by an expert, and in a relatively small number due to the strong bias introduced by the representation change.

This work has numerous perspectives. The first one is the issue of scaling up, in order to handle more complex data, such as the very recent array-CGH which measures genomic alterations at 30,000 different *BACs* instead of about 3,000 [17] SNP-arrays, or gene expression micro-arrays. This may lead us to relax the completeness dogma we have adopted until now. In that case, we may investigate the sequential sampling strategy [29]. Another way to extend this work would be to enrich the language of local patterns to represent additional information concerning the samples, or the function of genes included in the regions or close to the regions. This seems a promising way to augment the recall and precision of the discriminant patterns already obtained.

Acknowledgements This research has been made possible thanks to close collaborations with D. Chopin group (EMI 0337, INSERM, Hôpital Henri Mondor, Créteil), with D. Pinkel (Cancer Center, UCSF, USA) and with the Bioinformatics group of

Institut Curie, supervised by E. Barillot. We wish to thank Ph. Hupé for the *GLAD* software, Ph. La Rosa for the visualisation interface for CGH data, MAIA. Thanks to N. Stransky for his many suggestions concerning CGH data manipulation, to W. Wynant for his advice on the statistical evaluation of the obtained rules and to A.V. Salle for her support in data preparation. This work has been partially supported by the european project HKIS IST-2001-38153.

References

1. D.G. Albertson, C. Collins, F. McCormick, and J.W. Gray. Chromosome aberrations in solid tumors. *Nat Genet*, 34:369–76, 2003.
2. C.F. Aliferis, D. Hardin, and P.P. Massion. Machine learning models for lung cancer classification using array comparative genomic hybridization. In *Proc AMIA Symp. 2002*, pages 7–11, 2002.
3. R.J. Bayardo, R. Agrawal, and D. Gunopulos. Constraint-based rule mining in large, dense databases. *Data Mining and Knowledge Discovery*, 4:217–240, 2000.
4. C. Becquet, S. Blachon, B. Jeudy, J.F. Boulicaut, and O. Gandrillon. Strong association rules mining for large gene expression data analysis : a case study on human SAGE data. *Genome Biology*, 12, 2002.
5. J. Besson, C. Robardet, J.F. Boulicaut, and S. Rome. Constraint-based concept mining and its application to microarray data analysis. *Intelligent Data Analysis*, 9. to appear.
6. C. Billerey, D. Chopin, M.H. Aubriot-Lorton, D. Ricol, S. Gil Diez de Medina, B. Van Rhijn, M.P. Bralet, M.A. Lefrère-Belda, J.B. Lahaye, C.C. Abbou, J. Bonaventure, E.S. Zafrani, T. Van der Kwast, J.P. Thiery, and F. Radvanyi. Frequent FGFR3 mutations in papillary non-invasive bladder (pTa) tumors. *Am. J. Pathol.*, 158:1955–1959, 2001.
7. F. Bonchi, F. Giannotti, A. Mazzanti, and D. Pedreschi. Examiner: Optimized level-wise frequent pattern mining with monotone constraints. In *Proc. of the Third IEEE Int. Conf. on Data Mining (ICDM'03)*, 2003.
8. F. Bonchi and C. Lucchese. On closed constrained frequent pattern mining. In *Proc of the Fourth International Conference on Data Mining (ICDM'04)*. Morgan Kaufmann, 2004.
9. E. Boros, V. Gurvich, L. Khachiyan, and K. Makino. On the complexity of generating maximal frequent and minimal infrequent sets. In *Symposium on Theoretical Aspects of Computer Science*, pages 133–141, 2002.
10. C. Bucila, J. Gehrke, D. Kifer, and W. White. Dualminer: A dual-pruning algorithm for itemsets with constraints. *Data Mining and Knowledge Discovery*, 7:241–272, 2003.
11. D. Cappellen, S. Gil Diez de Medina, D. Chopin, JP. Thiery, and F. Radvanyi. Frequent loss of heterozygosity on chromosome 10q in muscle-invasive transitional cell carcinomas of the bladder. *Oncogene*, 14:3059–66, 1997.
12. K. Chin, C. Ortiz de Solorzano, D. Knowles, A. Jones, W. Chou, E. Garcia Rodriguez, W.L. Kuo, B.M. Ljung, K. Chew, K. Myambo, M. Miranda, S. Krig, J. Garbe, M. Stampfer, P. Yaswen, J.W. Gray, and S.J. Lockett. In situ analyses of genome instability in breast cancer. *Nat Genet*, 36:984–988, 2004.
13. A. Clare and R.D. King. Predicting gene function in saccharomyces cerevisiae. *Bioinformatics*, 19 Suppl. 2:ii42–ii49, 2003.

14. G. Dong and J. Li. Efficient mining of emerging patterns: Discovering trends and differences. In *Knowledge Discovery and Data Mining*, pages 43–52, 1999.
15. W. C Hahn and R.A. Weinberg. Modelling the molecular circuitry of cancer. *Nature*, 2:331–341, 2002.
16. P. Hupé, N. Stransky, J.P. Thiery, F. Radvanyi, and E. Barillot. Analysis of array CGH data: from signal ratio to gain and loss of DNA regions. *Bioinformatics*, 12:3413–22, 2004.
17. A.S. Ishkanian, C.A. Malloff, S.K. Watson, R.J. DeLeeuw, B. Chi, B.P. Coe, A. Snijders, D.G. Albertson, D. Pinkel, M.A. Marra, V. Ling, C. MacAulay, and W.L. Lam. A tiling resolution DNA microarray with complete coverage of the human genome. *Nat Genet*, 36:299–303, 2004.
18. S. Kramer and L. De Raedt. Feature construction with version spaces for biochemical application. In *Proc of the 18th International Conference on Machine Learning (ICML-2001)*. Morgan Kaufmann, 2001.
19. J. Li, G. Dong, K. Ramamoharao, and L. Wong. DeEPs: a new instance-based discovery and classification system. *Machine Learning*, 54:99 – 124, 2004.
20. J. Li and L. Wong. Identifying good diagnostic gene groups from gene expression profiles using the concept of emerging patterns. *Bioinformatics*, 18:725–734, 2003.
21. G. Mercier, N. Berthault, J. Mary, J. Peyre, A. Antoniadis, J.P. Comet, A. Cornuejols, C. Froidevaux, and M. Dutreix. Biological detection of low radiation doses by combining results of two microarray analysis methods. *Nucleic Acids Res.*, 32, 2004.
22. F. Pang, G. Cong, A.K.H Tung, J. Yang, and M. Zaki. Carpenter : Finding closed patterns in long biological datasets. In *Proc of SIGKDD'03*, 2003.
23. N. Pasquier, Y. Bastide, R. Taouil, and L. Lakhal. Discovering frequent closed itemsets for association rules. In *7th Intl. Conf. on Database Theory*, pages 398–416, 1999.
24. J. Pei, J. Han, and R. Mao. Closet an efficient algorithm for mining frequent closed itemsets. In *Proc. of the ACM SIGMOD Workshop on Research Issues in Data Mining and Knowledge Discovery (DMKD)*, 2000.
25. J.R. Quinlan. *C4.5: Programs for machine learning*. Morgan Kaufman, 1993.
26. F. Rioult, J.F. Boulicaut, B. Crémilleux, and J. Besson. Using transposition for pattern discovery from microarray data. In *Proc. of the 8th ACM SIGMOD Workshop on Research Issues in Data Mining and Knowledge Discovery (DMKD03)*, pages 73–79, 2003.
27. C. Rouveirol et al. Computation of minimal recurrent gain and loss regions from array-CGH data. Extended version of JOBIM2004, in preparation.
28. L. Saitta and J.D. Zucker. Semantic abstraction for concept representation and learning. In *Symposium on Abstraction, Reformulation and Approximation (SARA98)*, pages 103–120, 1998.
29. T. Scheffer and S. Wrobel. Finding the most interesting patterns in a database quickly by using sequential sampling. *Journal of Machine Learning Research*, 3:833–862, 2002.
30. A.M. Snijders, N. Nowak, R. Segraves, S. Blackwood, N. Brown, J. Conroy, G. Hamilton, A.K. Hindle, B. Huey, K. Kimura, S. Law, K. Myambo, J. Palmer, B. Ylstra, Y.P. Yue, J.W. Gray, A.N. Jain, D. Pinkel, and D.G. Albertson. Assembly of microarrays for genome-wide measurement of DNA copy number. *Nat Genet*, 29:263–4, 2001.
31. A. Soulet, B. Crémilleux, and F. Rioult. Condensed representation of emerging patterns. In *Proc. PAKDD 2004*, pages 127–132, 2004.

Appendix

smi_+	smi_-	sma_-	sma_+	corr	FP	FN	uncl.
5	5	0	0	24	2	5	32
5	5	1	1	39	1	5	18
5	10	0	0	24	2	5	32
5	10	1	2	41	0	5	17
5	15	0	0	22	3	4	34
5	15	1	2	41	1	4	17
5	20	0	0	19	3	2	39
5	20	1	2	40	1	3	19
10	5	0	0	24	2	5	32
10	10	1	2	41	0	5	17
10	15	0	0	22	3	4	34
10	15	1	2	41	1	4	17
10	20	0	0	19	3	2	39
10	20	1	2	40	1	3	19
15	5	0	0	24	0	5	34
15	5	2	1	39	1	5	18
15	10	0	0	24	0	5	34
15	10	1	2	43	0	5	15
15	15	0	0	23	1	4	35
15	15	1	2	42	1	4	16
15	20	0	0	20	1	2	40
15	20	1	2	41	1	3	18
20	5	0	0	24	0	5	34
20	5	1	1	40	1	5	17
20	10	0	0	24	0	5	34
20	10	1	2	43	0	5	15
20	15	0	0	23	1	4	35
20	15	1	2	42	1	4	16
20	20	0	0	20	1	2	40
20	20	1	2	41	1	3	18

smi_+	smi_-	sma_-	sma_+	corr	FP	FN	uncl.
5	5	0	0	25	1	9	38
5	5	0	1	31	1	10	31
5	10	0	0	10	2	8	53
5	10	0	1	30	1	9	33
5	15	0	0	8	2	5	58
5	15	0	1	22	1	9	41
10	5	0	0	26	0	9	38
10	5	0	1	31	0	10	32
10	10	0	1	30	0	9	34
10	10	0	0	11	0	8	54
10	15	0	0	9	0	5	59
10	15	0	1	23	0	9	41

Fig. 7. Excerpt of leave-one-out validation results for the class *FGFR3 mutated/non mutated* (left) and *invasive/non invasives* (right). For each pair $smin_+$ and $smin_-$, we provide the values for $smax_-$ and $smax_+$ which give the best and worst predictions . *corr*: number of corrected classified examples, *FP*: number of false positives, *FN*: number of false negatives, *uncl.*: number of examples which are be classified (tie)

Learning with Local Models

Stefan Rüping

University of Dortmund, 44221 Dortmund, Germany,
rueping@ls8.cs.uni-dortmund.de,
http://www-ai.cs.uni-dortmund.de/

Abstract. Next to prediction accuracy, the interpretability of models is one of the fundamental criteria for machine learning algorithms. While high accuracy learners have intensively been explored, interpretability still poses a difficult problem, largely because it can hardly be formalized in a general way. To circumvent this problem, one can often find a model in a hypothesis space that the user regards as understandable or minimize a user-defined measure of complexity, such that the obtained model describes the essential part of the data. To find interesting parts of the data, unsupervised learning has defined the task of detecting local patterns and subgroup discovery. In this paper, the problem of detecting local classification models is formalized. A multi-classifier algorithm is presented that finds a global model that essentially describes the data, can be used with almost any kind of base learner and still provides an interpretable combined model.

1 Introduction

It is commonplace knowledge that more and more data is collected everywhere and that the size of data sets available for knowledge discovery is increasing steadily. On the one hand this is good, because learning with high-dimensional data and complex dependencies needs a large number of examples to obtain accurate results. On the other hand, there are several learning problems which cannot be thoroughly solved by simply applying a standard learning algorithm to all the available examples. While the accuracy of the learner typically increases with example size, other criteria are negatively affected by too much examples. This paper will deal with the criterion of interpretability of the learned model, which is an important, yet often overlooked aspect for applying machine learning algorithms to real-world tasks.

The key problem with interpretability is that humans are very limited in the level of complexity they can intuitively understand [10]. An optimal solution of a high-dimensional, large-scale learning task, however, may lead to a very large level of complexity in the optimal solution. What can we do about this problem? Experience shows that one can often find a simple model which provides not an optimal solution, but a reasonably good approximation. The hard work usually lies in improving an already good model. Hence, we can try to find a simple model first and then concentrate on improving only those parts of the input

K. Morik et al. (Eds.): Local Pattern Detection, LNAI 3539, pp. 153–170, 2005.
© Springer-Verlag Berlin Heidelberg 2005

space, where the model is not good enough. This will be an easier task because less examples have to be considered and hence one might use a more sophisticated learner. In other words, one constructs not one single global model for all the data, but a global model plus one or more local models to cover special cases. Also, for the aspect of discovering new knowledge, it may happen that the global model finds only the obvious patterns in the data that domain experts are already aware of. Patterns are more informative, if they are surprising [8], i. e. if they contradict what is already known. Hence, it may also be the case that the local models actually contains the interesting cases.

To avoid problems with finding general measures of complexity for arbitrary models, this paper takes a very hands-on approach to interpretability: The user is allowed to use an arbitrary global learner, such that he can pick the learner whose results are the most understandable to him. He is also allowed to use an arbitrary clustering algorithm, which will be used to identify where to use the global algorithm and where not. So he can use the clusterer which will give him the most understandable description, where his global algorithm is right. To improve accuracy, a high-performance local learner is used to predict the rest of the examples, as long as the deviation from the global model remains below a user-defined threshold.

Summing up, the goal of this paper is to develop an hierachical, multi-classifier algorithm for learning local models which

1. learns a global model that sums up the essential properties of the data
2. can use (almost) every kind of learner to find the global and local base models
3. reduces a given criterion of model complexity for the global model.

The rest of the paper is organized as follows: The next section will discuss a broader picture of local models, complexity and interpretability. Readers that are only interested in the proposed local model algorithm may skip this section. Section 3 will review some learning algorithms which will be needed as parts of our hierachical approach. The new local model algorithm will be presented in Section 4, while Section 5 will discuss related approaches. Experimental results are given in Section 6 and Section 7 gives some conclusions.

2   Local Models

There are two aspects to local models: structure and performance. The structural aspect refers to the case where the optimal model of the data is composed of several parts that have a meaningful interpretation in terms of the application. This may be the case because the structure behind the data can only be expressed in a combination of hypothesis spaces of several standard learners (e. g. a combination of numerical and logical rules) or because there are limits in terms of interpretability. The performance aspect of local models refers to the case that a good model can theoretically be found by a single learner on the whole example set, but it is more efficient to use separate learning runs. Here, the distinction between global and local models is only meaningful in terms of algorithmic complexity and time and space requirements, but not in any intrinsic way.

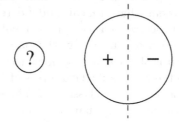

Fig. 1. Relationship between local patterns and local models. The left circle is a local pattern, but only a local model if its class is negative.

There is an obvious connection between local models and the detection of local patterns [9]. Local pattern or subgroup detection is defined as the unsupervised detection of high-density regions in the data. There are several way to formalize local patterns. In the scope of this paper, local patterns are defined as follows:

Definition: Given an input space X and some default probability measure $P_{Default}(X)$, a local pattern is a subset X' of X such that the empirical probability[1] of X' is significantly different from $P_{Default}(X')$.

The idea is that the user already has some idea of what his data looks like (for example he assumes that all items in a store are bought independently of one another) and he his interested in cases where his beliefs are wrong (for example an association rule describing which items are often bought together). In contrast, local models are meant to improve the classification, hence the interesting quantity is the class probability $P(y|x)$.

Definition: Given an input space $X \times Y$ and a default (or global) class probability measure $P_{Default}(Y|X)$, a local model is a subset X' of X such that the empirical probability $P_{emp}(Y|X')$ is significantly different from $P_{Default}(Y|X')$, plus a classification rule with input space X'.

Here, the user will not be bothered with deviations from the data to his beliefs, as long as this does not have an influence on the attribute of interest y. The difference can be seen in Figure 1. In a local pattern task, the default probability could be a Gaussian approximating the large batch of data in the circle on the right and the corresponding local model would be a smaller Gaussian describing

---

[1] The empirical probability of X' is the frequency of X' observed in the data, in contrast to the expected frequency as defined by $P(X')$.

the circle on the left. In a local classification task, an obvious default model would be the vertical line through the large circle, classifying all points to its right as negative and the points to its left as positive. In contrast to the local pattern case, the smaller circle will only be a local model if its class is negative, i. e. different from the one predicted by the default model. Summing up, the difference between both tasks is that in local model detection the goal is not to just find regions with unsuspected density, but to identify those which can be used to improve the overall performance, e. g. the classification accuracy.

Given these definitions it is obvious that the local model task can be solved by detecting local patterns in the misclassified examples of the default classification rule. This approach learns a very simple classification model: predict the negative of the default rule. Its disadvantage is, that the simplicity of the classifier may lead to the problem of learning very complex patterns. So the complexity of the classification is not removed, but only transferred to the decision, where to apply the classifier, i. e. the local patterns. As the classification problem is usually much better investigated in machine learning than the local pattern problem, it will usually be better to define a more simple pattern together with a more complex decision function, see Figure 2. We will take this approach in this paper.

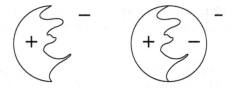

Fig. 2. Given the surrounding data is negative, the positive data can be marked in two ways: a complex cluster plus a trivial classifier (default positive, right) versus a trivial cluster (circle) plus a complex classifier (twisted line, right).

From the perspective of interpretability, the use of simple local patterns together with a more complex local classifier allows to better separate the local and global models, because it minimizes the interactions between both models. It allows to interpret both models on its own without the need to take interactions into account, which account for a large part of the complexity of the overall classifier.

Talking about interpretable models, one has to acknowledge that the concept of interpretability is very hard to formalize. Interpretability of models is often evaluated by interviewing a human expert, but reliable information is hard to come by, as this would need a survey of several independent human experts. Findings from psychology show that the size of the models plays some part, as humans are usually only able to deal with about seven cognitive units at the same time [10]. This motivates certain measures of model complexity like the number or length of rules in logical models [18], which are traditionally assumed to be more interpretable than other approaches, as logical rules can be easily cast into sentences in natural language. However, experience shows that the experience of humans with interpreting the kind of model or its visualisation plays a crucial role. Plainly, humans tend to find the things the most interpretable, they are already acquainted with. Concludingly, this paper takes a very pragmatic approach to the question of interpretability: let the user choose whatever learning algorithm he likes best and optimize a user-given criterion of model complexity.

3   Learning Algorithms

In this section we will shortly introduce the learning algorithms that were used in the experiments in this paper. A detailed discussion of these algorithms is beyond the scope of this paper, we will restrict ourselves to those properties that are of interest for the course of this paper. The learning tasks to solve are both density estimation and probabilistic classification.

3.1   Density Estimation

Several algorithms exists for the approximation of a density $P(x)$ from examples (x_i). We limit this discussion to the approximation by a Gaussian as the prototypical parametric approach and probably easiest density estimation technique.

To approximate the data by a Gaussian distribution one has to notice that the multivariate Gaussian is completely determined by its mean and covariance matrix. Hence, it suffices to calculate the mean and covariance matrix of the data and substitute it into the Gaussian distribution. This is a very simple algorithm, but also a very unflexible, as it is limited to a very special parametric form. However, it turns out that this is sufficient for the local model algorithm of this paper, as we do not need to approximate the density excactly, but only need the densities to separate different local regions in the input space in order to split the classification model up into less complex parts. This approach is similar to the well-known k-means clustering algorithm, where one is not interested in the form and density of each cluster itself, but only in the border line between each cluster and the rest.

What is important in the course of this paper is the robustness of the density estimation. Robustness means that if a certain fraction of examples is not drawn from the distribution of interest, this should have little influence on the estimated density. Standard estimates of mean and covariance can be heavily distorted by far away outliers and hence the trivial Gaussian approximation is not robust. In this paper, we use the Minimum Covariance Determinant estimator [15], which searches for a subset of the examples of given size such that the determinant of its covariance matrix becomes minimal and hence, the examples are located very close together. The mean and covariance of this subset are then again used to define a Gaussian density.

3.2   Probabilistic Classification

The goal of probabilistic classification is to not only find a classifier, but also an estimation of the conditional class probability $P(y|x)$. A general, but very coarse solution is to use the classifiers accuracy or its precision for the positive and negative class, respectively.

Numerical classifiers, i. e. classifiers of the form $cl(x) = sign(f(x))$, can be transformed into probabilistic classifiers by finding an appropriate scaling function σ such that $P(Y = 1|x) = \sigma(f(x))$. This scaling function can be found by minimizing the cross-entropy

$$CRE = \sum_{i:y_i=1} \log(\sigma(f(x))) + \sum_{i:y_i=-1} \log(1 - \sigma(f(x)))$$

or the mean squared error

$$MSE = \frac{1}{n} \sum_i (\frac{1 + y_i}{2} - \sigma(f(x)))^2$$

over the space of applicable scaling functions.

In this paper Support Vector Machines [22] are used as numerical classifiers. For Support Vector Machines, appropriate scaling procedures have been investigated in [12,16]. A generic scaling function for a large variety of learners has also been proposed in [7]. Other classifiers can directly give an estimate of the conditional class probability, for example decision trees [14] where the class distribution at each leaf or its laplacian correction can be used as an estimator of the class probability of an observation classified by this leaf.

3.3   Interpretable Learners

One of the goals of this paper is the interpretability of the classifier. As was already pointed out in the introduction, the user may in principle prefer any kind of classifier. In the following, we limit the discussion to decision trees as interpretable learners, because they have some favourable properties with respect to interpretability: First, decision trees can be easily visualised, because they consist of a simple tree structure and simple tests. They can alse be transformed into a set of independent rules, which the user can investigate one after the other. Secondly, there are two simple measures of model complexity, namely the depth of the tree and its number of nodes. It is obvious that a decision tree is the more understandable, the less tests have to be made in order to classify an observation and the less tests it contains in general. This allows us to quantify the degree of complexity reduction. Thirdly, these measures of complexity can directly be optimized at the construction of the tree by cutting the tree off at the maximal allowed depth and continuing with the pruning phase of the tree induction until the maximum number of nodes is met.

3.4   Expectation Maximization

The Expectation Maximization (EM) algorithm [2] is a general technique for computing the maximum likelihood estimates of the parameters of a distribution in the presence of hidden variables. This approach assumes that in addition to the observed data X, there are hidden variables Z, such that the observations (x_i) could be modeled much better if the (z_i) were known. The EM algorithm involves two steps, the expectation step and the maximization step. The E-step computes the values of the hidden variable z_i, or a sufficient description thereof, given the current estimate of the parameters. The M-step computes the parameters as the maximum likelihood estimation given the observed data and the current estimate of the hidden variables. Both steps are iterated until convergences or a sufficient number of times. It can be shown that the EM algorithm converges to a local optimum under some very general assumptions. The well-known k-means clustering algorithm is a famous application of the expectation maximization algorithm.

4   Algorithm

Following the earlier discussion on complexity and interpretability, we assume the following problem setting: We are given data $(x_i, y_i), i = 1 \ldots n$, two arbitrary learners $\mathcal{L}_{Glob}$ and $\mathcal{L}_{Loc}$, a density estimation algorithm and a real number $\tau \in$ $]0, 1[$. We want to find a model that is a combination of a global model learned from $\mathcal{L}_{Glob}$ and several local models learned from $\mathcal{L}_{Loc}$, such that the combined model differs from the global model by at most τ. More formally, we define dummy variable $z_i, i \in 1 \ldots k$ for the single base models, and let the combined model take the form

$$P(Y = 1|x) = \frac{1}{P(x)} \sum_{j=1}^{k} P(Y = 1|x, z = j)P(x|z = j)P(z = j),$$

where

$$P(x) = \sum_{j=1}^{k} P(x|z = j)P(z = j)$$

with $P(z = 1) > 1 - \tau$. We replaced the classifiers $y = f(x)$ by estimators of the conditional class probability $P(Y = 1|x)$, which can be done by one of the algorithms described in Section 3.2. Hence, by inspecting $P(Y = 1|x, z = 1)$ (the global model), the user learns how the combined model behaves on a fraction of $P(z = 1) > 1 - \tau$ of the cases, while inspecting the $P(x|z = j)$ tell him where the global model is applied and where not. Note that although at each point x the combined prediction is a linear combination of the base models predictions, the combined model is a nonlinear combination, as the $P(x|z = j)$ are not linear.

The local model problem as it is defined here is a most general black-box setting, as we neither assume knowledge about the internals of the learning and density estimation algorithms, nor about the interpretation of its models. In particular, we do not assume that it is possible to assign weights to example, modify the learners parameters or deduce the influence of specific training examples to the learners model. The only possibility for the overall algorithm to interact with the given learners is to present the learners different training sets. For the outputs, we only assume that the classifiers return a real value and that the higher this value is, the more certain the classifier is that the observation belongs to the positive class (in particular, this includes the case of a simple binary classifier $f(x) \in \{-1, +1\}$). It was shown by Garczarek [7] that this is sufficient to convert the classifiers output into an estimate of the conditional class probability $P(Y = 1|x)$.

Equation 4 contains two special cases for learning interpretable models as extremes: For $P(z = 1) = 1$, we use only the global learner. This gives the user full control over what the learner does, but only in a few lucky cases the most understandable and the most accurate model will coincide. For $P(z = 1) = 0$, one may use the most accurate model. The user may still inspect a model from the global learner to understand the data, but he has no control over in which cases the interpretable and the accurate will disagree, as he is missing the explicit model $P(x|z = i)$ of global and local regions.

Of course, there are many other methods for combining several learners and one might ask, whether explicitely learning density models in the mixture approach as described in Equation 4 is not a more complex task than necessary. To justify the mixture idea, let us compare it with two simpler approaches: First, in Section 3.2 we described the idea of probabilitically scaling a classifier to obtain an estimate of the conditional class probability $P(y|x)$. Now, one might think of using the most confident classifier for each example or to combine the confidences of each classifier. But this idea is fallacious, because in general each classifier can only be trusted to give good probability estimates over the region of the input space it was trained and scaled on. For example, a linear classifier will be the more confident the further away from the decision line the observations lie, regardless of the position of its training examples. Hence, a local classifier may either give much too optimistic results on non-local examples (when scaled over the local examples only) or too pessimistic results on the local examples (when scaled over many examples it was not trained on).

A second seemingly easier approach to learning local models would be to use the local learner to predict whether the global learner is right or not. But this, too, is not a good approach, because it makes the local learning problem much more complex. The learner has not only to find structures in the data but also the structure superimposed by the global model. Besides from complicating the learning problem, this effectively prohibits to understand the overall prediction by looking at the global model, as any prediction from the global model might be negated by the second model. Hence, in this paper we use a combination of models were each learner is trained to directly predict the true class.

4.1   Learning the Combined Model

To find the combined model, we borrow ideas from two well-known learning algorithms: classification with covering algorithms and EM clustering. Covering algorithm find a logical rule which covers a part of the data, remove the examples covered by this rule and then iteratively find further rules to cover the rest of the data. Because of their iterative nature, they are well suited for the task of finding local models, when one views the first rule as the global rule and the following rules as local models [5] Learning by covering makes explicit use of the fact that logical rules (e. g. "$X1 = a \wedge X2 \leq b \Rightarrow Y = 1$") make predictions for only a part of the input space, such that it is clear which rule can be applied. But this means that this idea cannot be directly applied to other base learners, as in general, learners may make a prediction for every observation in the input space. This is the reason why we need an additional density estimator to select which model to use.

To find both the models and the clusters, on which each model should be applied, we may proceed similar to EM clustering. EM clustering algorithms iteratively find a cluster model, then for each example estimate the probability of belonging to each of the clusters and re-estimate the clusters using these probabilities as weights (i. e. an example with low probability of belonging to some cluster will have little influence on the shape of this cluster). One might be tempted to solve the local model problem by directly using clustering as a pre-processing step, perhaps with a minimum size constraint on the first cluster, and then find a different local model for each cluster. But this strategy may fail, as clustering groups observations according to some similarity measure in X, while we are interested in grouping observations together if they can be predicted by the same model. If observations look very different in the input space, but can be correctly predicted by the same simple model, there is no reason to treat them differently as far as classification is concerned. This is the same situation as described in Figure 1.

The reason for using an EM-like approach instead of a greedy approach as in covering algorithms is that even if there is a simple structure behind parts of the data that one learner could find, the learners model may be distorted by outliers from the rest of the data. This problem can be seen in Figure 3 in Section 6, where a linear hyperplane from a Support Vector Machine is distorted by a small set of far away outliers. Once the local examples are known as such, it is very easy to improve the model by removing this examples from the training set of the classifier. Hence, it is important to re-evaluate the models once more information about clusters and outliers is present. The Expectation Maximization procedure allows to optimize both models in parallel by finding an optimal allocation from examples to learners.

As we are interested in predicting x, i. e. in the conditional class distribution $P(y|x)$, in the final application of our model only x, but not z, is known, such that we can make use of the mixture decomposition 4 of $P(Y = 1|x)$ only if z can be identified from x alone. Hence, we need the following assumption:

Assumption: The distributions $P(x|z = j)$ differ significantly for two different j_1, j_2.

We will not define formally, what a significant difference of two probability distributions is, as this is not crucial for the algorithm presented here. What is important is the intuition that the decomposition in k distributions will only be of benefit if the distributions live in different part of the input space X. Accordingly, in this paper the terms *j-th probability distribution* $P_j(x)$, *j-th cluster* and *j-th batch of data* are used interchangeably. It is easy to check whether this assumption holds by computing

$$P(Z = j|x) = \frac{1}{P(X)} \sum_i P(x|Z = j)P(Z = j)$$

for each x in the trainig set. If the assumption holds, the distribution of $P(Z = j|x)$ should follow an U-form, i. e. most examples should either clearly belong to the j-th cluster ($P(Z = j|x) \approx 1$) or clearly belong to a different cluster ($P(Z = j|x) \approx 0$).

In a similar way to $P(z|x)$, we can also compute $P(z|x, y)$ as

$$P(z|x, y) = \frac{P(x, y, z)}{P(x, y)} = \frac{P(x, y, z)}{\sum_z P(x, y, z)}$$

with

$$P(x, y, z) = P(y|x, z)P(x|z)P(z).$$

Note that here we need to transform the conditional class estimate $P(Y = 1|x, z)$ given by the learner into the probability $P(y|x, z)$ that the learner predicts the correct class. Following the update of $P(z|x, y)$, the default probability $P(z)$ can be re-estimated as $P(z) = avg_i P(z_i|x_i, y_i)$.

Theoretically, we could now follow the algorithm for EM clustering algorithm by using $P(z|x, y)$ to assign examples to clusters and then use the learner and density estimator on each cluster to learn an update of $P(y|x, z)$ and $P(x|z)$. But it turns out that we need another intermediate step. The problem is the allocation of examples to batches in the presence of noise. Imagine the case when a large part of the error is due to random noise in y, independent of x, e. g. by independently flipping any label with a fixed, small probability. When we have an approximately correct global model, every non-flipped example will have a high $P(z|x, y)$, while every flipped example will have a very small probability belonging the global model, as the model can be quite certain that the example belongs to its cluster (high $P(x)$) and can be quite certain that its prediction is correct on the average (high $P(y|x)$), but the prediction indeed is wrong (flipped label). As a result, the examples with the lowest probability $P(z|x, y)$ will most likely be the errors of the global model and hence, the best local classifier to learn from these examples is the negative of the original classifier. Now one would need to know if the first classifier is wrong beforehand, because by the independence of $P(noise)$ and $P(x)$, the observation x is completely uninformative to the correctness of the first classifier. This is of course totally useless.

To remedy this problem we double the EM-step of the algorithm: in the first step, we allocate examples to batches with respect to $P(z|x, y)$ (E-step 1) and then learn a cluster model $P_j(x) = P(x|z)$ only (M-step 1). In the second step, we allocate examples to batches with respect to the learned $P(z|x)$ (E-step 2) and conclude with learning the classifier $P_j(x) = P(y|x, z)$ from the new batches (M-step 2). The trick is that now in the first M-step the density estimator only sees the examples that this model can predict better than any other, while in the second M-step the classifier does see all the examples it will be asked to predict later. That is, the first step is to find a local pattern (deviation from the combination of the rest of the models), and the second step is to learn a model for exactly this pattern.

In each E-step, we take care to partition the examples into disjunct batches for each learner by first choosing the examples for the batches with higher $p(z)$ and then choosing the examples for batches with lower $p(z)$ from the rest. This makes sure that different batches do not learn redundant models.

Alternatively, one could also give each learner the complete example set plus weights based on $P(z|x)$ or $P(z|x, y)$, as in standard k-means. We do not use this approach here, because we want to be able to 'plug-in' as many different learners as possible and there exist many types of learners that cannot deal with example weights. A possible alternative to the approach defined here is to not partition the example set but allow the batches to overlap themselves to a certain degree, in order to account for undecisive batch probabilities.

4.2   Finally, the Algorithm

The final algorithm looks like this:

```
Local Model Algorithm:
1  input: data (xi,yi), #clusters k, #iterations i,
           threshold tau
2  independently learn k models Pj(X,Y)
3  repeat i times
4      for all j estimate P(zi=j|xi,yi) from Pj
5      adjust P(zi=j|xi,yi) such that P(z=1) >= 1-tau
6      assign observations (xi) to batch j = argmax(P(zi=j|xi,yi))
7      for all j estimate P(zi=j|xi) from Pj
8      adjust P(zi=j|xi) such that P(z=1) >= 1-tau
9      assign examples (xi,yi) to batch j = argmax(P(zi=j|xi))
10     for all j learn model Pj(x,y) from batch j
```

If $P(z = 1) < \tau$ in step 5 or 8, we set $P(z = 1) = \tau$ and adjust all other $P(z = j)$ linearly.

As we do not use weights for the examples, but a hard threshold to decide whether or not to include an example in a training set, we cannot give a convergence result as in standard k-means or other EM algorithms (in k-means, the models are continuous in the weights of the examples, but of course there is no continuity in including or excluding an example). Hence, the final model is chosen as the model with minimal training error.

5 Related Work

There exists several approaches for combining multiple classifiers, for example Voting, combination by order statistics [21], Meta-Level Learning [1], Stacking [23], Cascade Generalization [6] and Boosting [3]. In Boosting, the combined classifier is a linear combination of the base classifiers. The single classifiers and their weights are learned iteratively. In each step, explicit information about the error of the combined classifier so far is used and the classifier is added, that reduces the error of the combined classifier the most in terms of a certain loss function [4]. This idea is implemented by assigning a weight to each example. After each step, the weights of the correctly classified examples are reduced an the weights of the misclassified examples are increased. The weights can be seen as a probability distribution over misclassifications.

Boosting is a most successful approach in terms of accuracy, but the interpretability of its model is very limited. Following the terminology from this paper, one might be tempted to call the first base model the global model and the following models, that are learned with respect to the error distribution of the combined model so far, local models. The problem is, that the following models are not meaningful by themselves, but only with respect to all the models and their weights learned so far. If for example the i-th model shows that a certain combination of attribute values is indicative of the positive class, this does not mean that there is a correlation between these attribute values and the positive class in the data, but only means that the combined classifier so far has for some reason estimated too much influence of these attributes to the negative class. Also, Boosting is a greedy combined learner, i. e. previous models are not corrected once they have been learned, even if it turns out that they are wrong in several parts.

With respect to interpretability, most other combined learners suffer from the same problem, namely that to understand the model one has to understand every single base model plus the way these models are combined. Even if the base models are trivial, their combination can be quite complex. Boosting with decision tree stumps is an example of that.

An understandable combination of classifiers needs some kind of orthogonality, such that the effect of one model is independent of the effect of the other models, to ensure that the problem can be validly split up into smaller independent parts. One way to ensure this orthogonality is to split up the input space and find out which classifier works best in the different regions. Splitting up the input space can be done either beforehand by clustering or inside the learning procedure. Examples of this approach are [19] and [20]. Decision trees also iteratively split up the input space, such that theoretically one could define the first levels of the tree as a partition of the input space and the following levels as separate classifiers for each partition (but this is probably stretching out the idea of local classifiers too far). More advanced, in [17] decision trees and kernel density estimators have been combined to smoothen the posterior class probabilities.

However, in most cases existing approaches are usually either not easily interpretable or limited to a specific class of base learners. The goal of this paper was to find an algorithm that keeps up interpretation and works with arbitrary base classifiers.

6   Experiments

Let us first investigate the proposed local model algorithm on an artificial data set. Figure 3 shows a 2-dimensional data set of 200 observations consisting of two Gaussians, centered at $(0, 0)$ and $(2, 2)$, respectively. The first batch contains 95% of the observations with a standard deviation of $\sigma = 1$, while the second batch is smaller both in term of number of observations (5%) and in standard deviation ($\sigma = 0.1$). The positive examples are the examples from the first batch with negative second coordinate plus the examples from the second batch. An additional error of 5% in the labels was randomly added.

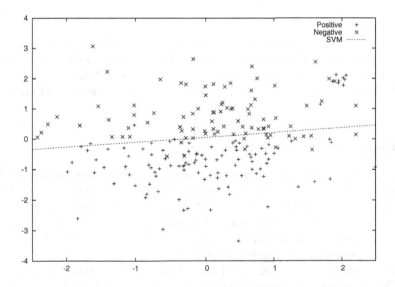

Fig. 3. Global model learned by a linear Support Vector Machine on a simple data set.

The straight line in Figure 3 shows a linear SVM classifier learned over all examples. One can see how the linear hyperplane is pulled into an ascending slope by the positive examples from the second batch.

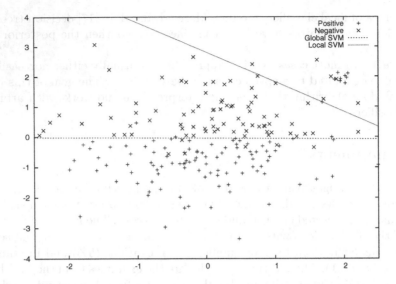

Fig. 4. Global model plus local model learned by the proposed algorithm with linear SVMs as the base learner.

Figure 4 shows the result of the local model algorithm after one iteration. In addition to the global linear classifier, a local linear classifier was added to classify the examples from the second batch as positive. Notice that the global classifier has gone back to the axis-parallel alignment that is optimal for the first batch.

6.1   Complexity Measure Reduction

The effect of interpretability improvement is of course hard to measure. To obtain quantifiable results, a C4.5 [13] decision tree learner is used in the following experiments, such that *interpretability* can be reduced to meaning *having a small number of nodes or levels*. To ensure an interpretable tree, the learning algorithm was modified such that the maximum depth of the tree was cut off at 75% the depth of the tree from the vanilla algorithm.

The experiments were conducted on 7 data sets, including 5 data sets from the UCI Repository [11] (breast, covtype, diabetes, ionosphere, liver) and 2 other real-world data sets: a business cycle analysis problem (business) and intensive care patient monitoring data (medicine). Prior to learning, nominal attributes were binarised and the attributes were scaled to expectancy 0 and variance 1. Multi-class-problems were converted to two-class problems by arbitrarily selecting two of the classes (covtype) or combining smaller classes into a single class (business, medicine). For the covtype data set, a 1% sample was drawn. The following table sums up the description of the data sets:

Name	Size	Dimension
breast	683	10
covtype	4951	48
diabetes	768	8
ionosphere	351	34
liver	345	6
business	157	13
medicine	6610	18

In addition to depth and number of nodes of the decision tree, the error of the combined classifier *C-error*, the error of the global decision tree classifier *G-error* and the disagreement between global and combined classifier *disagree*, i. e. the fraction of examples predicted differently by both classifiers have been recorded. All numbers reported are results of a 5-fold cross-validation.

The Support Vector Machine was used as local classifier. The type of kernel function (linear or radial basis) and the kernel parameters were selected beforehand by optimization over all examples. The C parameter was set at a default value. For density estimation, k-medoids, a robust version of k-means, was used.

In all experiments, one local model was learned and a fraction of $\tau = 0.3$ examples were allowed for this local model. 3 EM iterations were performed. The following table sums up the performance of the local model algorithm:

Data	Iteration	Depth	Nodes	C-Error	G-Error	Disagree
breast	1	5.0	17.4	0.042	0.042	0.0
	2	2.8	10.6	0.236	0.074	0.226
	3	1.2	3.4	0.029	0.067	0.055
covtype	1	22.4	528.2	0.228	0.229	0.001
	2	15.6	332.6	0.227	0.227	0.054
	3	15.6	366.6	0.227	0.239	0.058
diabetes	1	6.8	24.6	0.266	0.276	0.016
	2	4.0	14.6	0.242	0.250	0.251
	3	4.0	15.8	0.247	0.251	0.030
ionosphere	1	7.0	21.4	0.096	0.096	0.0
	2	4.0	9.4	0.116	0.119	0.054
	3	4.0	10.2	0.099	0.122	0.034
liver	1	9.8	49.0	0.313	0.318	0.005
	2	5.2	20.2	0.368	0.339	0.220
	3	6.0	24.6	0.350	0.347	0.205
business	1	6.4	20.2	0.223	0.223	0.0
	2	3.6	10.2	0.210	0.255	0.094
	3	3.6	12.2	0.216	0.222	0.057
medicine	1	19.2	389.4	0.202	0.204	0.017
	2	13.6	245.4	0.206	0.215	0.073
	3	13.6	239.8	0.211	0.219	0.104

It can be seen that the complexity of the tree classifier is reduced dramatically compared to the tree from the usual C4.5 algorithm (the first step is done without cutting off the tree, hence the size of the vanilla C4.5 tree can be seen in iteration 1). However, this complexity reduction does not decrease the classification performance, with the exception of the liver data set. On the average, in the third iteration the error is reduced by 4% while the size of the decision tree is reduced by 46%. Global and combined classifier differ in 7% of the cases. This shows, that the local model algorithm can effectively find a much less complex approximation to the optimal model.

7   Conclusion and Future Work

Local models are the extension of local patterns to the supervised learning case. They provide a good way to improve the interpretability of a classifier by restricting the classifier to the essential parts of the model and leaving out patterns that it hardly can approximate. In this paper, a local model algorithm was presented that learns a global classifier plus local models to reduce complexity of the global model, ensure the prediction quality of the combined model and provide guarantees that combined and global model will differ only up to a user-specified degree. This allows the user to restrict his attention to the global model and still get valid information about the high-quality combined model.

Another important aspect of this approach is the interpretability of the local models. In applications global rules often express only trivial knowledge about the data, that the user is already aware of, while the significant exceptions of these rules are highly informative. This aspect of local models will be dealt with in future work.

Another open problem concerns the runtime of the algorithm. Each iteration requires to learn a density and a classification model for both the global learner and the local learners. While the local models hopefully do not pose a problem, as only a small part of the data is concerned, the global models require to deal with possibly very large data sets. Theoretically, learning with local models could alleviate this problem, as it allows room for errors of the global model, which can be dealt with by local models. This would allow to use sampling or a faster, less accurate learner for the bulk of the data. It remains to be investigated how well this works in practice.

Acknowledgments

Many thanks to Stefan Wrobel for the very helpful comments to this paper. The financial support of the Deutsche Forschungsgemeinschaft (Collaborative Research Center SFB 475, "Reduction of Complexity for Multivariate Data Structures") is gratefully acknowledged.

References

1. Philip K. Chan and Salvatore Stolfo. Experiments in multistrategy learning by meta-learning. In *Proceedings of the second international conference on informationand knowledge management*, pages 314–323, Washington, DC, 1993.
2. A. P. Dempster, N. M. Laird, and D. B. Rubin. Maximum-likelihood from incomplete data via the EM algorithm. *Journal of the Royal Statistical Society Ser. B*, 39:1–38, 1977.
3. Y Freund and R.E Schapire. Game theory, on-line prediction and boosting. In *Proceedings of the 9th Annual Conference on Computational Learnin g Theory*, pages 325–332, 1996.
4. Jerome Friedman, Trevor Hastie, and Robert Tibshirani. Additive logistic regression: A statistical view of boosting. Technical report, Departement of Statistics, Stanford University, Stanford, California 94305, July, 23 1998.
5. Johannes Fürnkranz. From local to global patterns: Evaluation issues in rule learning algorithms. In Katharina Morik, Jean-Francois Boulicaut, and Arno Siebes, editors, *Detecting Local Patterns*. Springer, 2005.
6. João Gama and Pavel Brazdil. Cascade generalization. *Machine Learning*, 41(3):315–343, 2000.
7. Ursula Garczarek. *Classification Rules in Standardized Partition Spaces*. PhD thesis, Universität Dortmund, 2002.
8. Isabelle Guyon, Nada Matic, and Vladimir Vapnik. Discovering informative patterns and data cleaning. In Usama M. Fayyad, Gregory Piatetsky-Shapiro, Padhraic Smyth, and Ramasamy Uthurusamy, editors, *Advances in Knowledge Discovery and Data Mining*, chapter 2, pages 181–204. AAAI Press/The MIT Press, Menlo Park, California, 1996.
9. David Hand. Pattern detection and discovery. In David Hand, Niall Adams, and Richard Bolton, editors, *Pattern Detection and Discovery*. Springer, 2002.
10. G. Miller. The magical number seven, plus or minus two: Some limits to our capacity for processing information. *Psychol Rev*, 63:81 – 97, 1956.
11. P. M. Murphy and D. W. Aha. UCI repository of machine learning databases, 1994.
12. John Platt. *Advances in Large Margin Classifiers*, chapter Probabilistic Outputs for Support Vector Machines and Comparisons to Regularized Likelihood Methods. MIT Press, 1999.
13. John Ross Quinlan. *C4.5: Programs for Machine Learning*. Machine Learning. Morgan Kaufmann, San Mateo, CA, 1993.
14. R.J. Quinlan. Induction of decision trees. *Machine Learning*, 1(1):81–106, 1986.
15. P.J. Rousseeuw. Least median of squares regression. *Journal of the American Statistical Association*, 79:871–880, 1984.
16. Stefan Rüping. A simple method for estimating conditional probabilities in SVMs. In A. Abecker, S. Bickel, U. Brefeld, I. Drost, N. Henze, O. Herden, M. Minor, T. Scheffer, L. Stojanovic, and S. Weibelzahl, editors, *LWA 2004 - Lernen - Wissensentdeckung - Adaptivität*. Humboldt-Universität Berlin, 2004.
17. Padhraic Smyth, Alexander Gray, and Usama M. Fayyad. Retrofitting decision tree classifiers using kernel density estimation. In *International Conference on Machine Learning*, pages 506–514, 1995.
18. Edgar Sommer. *Theory Restructering: A Perspective on Design & Maintenance of Knowledge Based Systems*. PhD thesis, University of Dortmund, 1996.

19. L Todorovski and S. Dzeroski. Combining multiple models with meta decision trees. In *Proceedings of the Fourth European Conference on Principles of Data Mining and Knowledge Discovery*, pages 54–64. Springer, 2000.
20. Ljupco Todorovski and Saso Dzeroski. Experiments in meta-level learning with ILP. In J. M. Zytkow and J. Rauch, editors, *Proceedings of third European Conference on Principles of data mining and knowledge discovery (PKDD-99)*, volume 1704, pages 98–106. Springer, 1999.
21. Kagan Tumer and Joydeep Ghosh. Order statistics combiners for neural classifiers. In *Proceedings of the World Congress on Neural Networks*, 1995.
22. V. Vapnik. *Statistical Learning Theory*. Wiley, Chichester, GB, 1998.
23. D. Wolpert. Stacked generalizations. *Neural Networks*, 5:241–259, 1992.

Knowledge-Based Sampling for Subgroup Discovery

Martin Scholz

Artificial Intelligence Group
Department of Computer Science
University of Dortmund, Germany
scholz@ls8.cs.uni-dortmund.de

Abstract. Subgroup discovery aims at finding interesting subsets of a classified example set that deviates from the overall distribution. The search is guided by a so-called utility function, trading the size of subsets (coverage) against their statistical unusualness. By choosing the utility function accordingly, subgroup discovery is well suited to find interesting rules with much smaller coverage and bias than possible with standard classifier induction algorithms. Smaller subsets can be considered local patterns, but this work uses yet another definition: According to this definition global patterns consist of all patterns reflecting the prior knowledge available to a learner, including all previously found patterns. All further unexpected regularities in the data are referred to as local patterns. To address local pattern mining in this scenario, an extension of subgroup discovery by the knowledge-based sampling approach to iterative model refinement is presented. It is a general, cheap way of incorporating prior probabilistic knowledge in arbitrary form into Data Mining algorithms addressing supervised learning tasks.

1 Introduction

The discipline of Knowledge Discovery in Databases (KDD) is about finding useful and novel patterns, hidden in huge amounts of real-world data. A common problem is that the applied Data Mining techniques primarily find "obvious" patterns which are already known to domain experts. In this work we distinguish between global and local patterns, paying special attention to mining the local ones.

The notion of patterns is central to KDD. This work assumes that for a given target variable the absence of any pattern is equivalent to its independence of all the other variables. If prior knowledge is available then the absence of further patterns means that the prior knowledge models the distribution of the target variable precisely. In turn, a pattern is defined as a regular deviation from the independence assumption or given prior model, respectively. Thus, it shows as a correlation between the given target attribute and the other variables that has not been reported yet.

As a first idea the reader might want to think of global patterns as those discovered easily, e.g. because of having a high correlation to the target attribute

K. Morik et al. (Eds.): Local Pattern Detection, LNAI 3539, pp. 171–189, 2005.

in a densely populated subset of the instance space. Whether these patterns reflect prior domain knowledge or are the result of an earlier application of a Data Mining technique, in any case we might be interested in finding further patterns. From a technical point of view, the presence of some patterns may increase necessary efforts to observe others. Given a sample of limited size a less frequent pattern showing little effect on the target attribute may easily be considered to be part of another pattern. Due to a lack of significance it may also be hard to distinguish such patterns from random noise. To this end a specific sampling technique is proposed, paying special attention to patterns of lower frequency in subsequent Data Mining iterations.

Defining local patterns is possible based on the learner's prior knowledge: deviation from expectation indicates the presence of patterns not yet discovered. These patterns are referred to as local patterns. For simplicity this work confines itself to probabilistic rules as the representation language for patterns. Guiding the discovery of patterns by unexpectedness is close to the idea of subgroup discovery, a learning task discussed in section 3 after some necessary definitions are given in section 2. As the main contribution of this work a generic sampling technique to incorporate prior knowledge into subgroup discovery is presented in section 4 and empirically evaluated in section 5.

2   Basic Definitions

This section embeds the problem of mining local patterns into a formal Data Mining framework. The notion of a pattern as used in this work is in-line with the definition given in [8]: A pattern is characterised as a subset of the instance space with an anomalously high local density of data points. Local patterns are defined in terms of a (global) background model. Probabilistic rules, for simplicity always predicting the value of a boolean target attribute, are considered to be our target representation language for local patterns. A definition of this formalism is given in subsection 2.2 after some more basic definitions.

2.1   Instance Space and Distribution

The two learning tasks discussed in this paper are subgroup discovery and classifier induction. Both tasks are supervised, so the learning step is performed based on a sample of classified examples. *Examples* are defined as elements of an *instance space* $\mathbf{X}$. Usually the instance space $\mathbf{X} = A_1 \times A_2 \times \ldots \times A_k$ is the Cartesian product of a fixed set of nominal and/or numerical attributes. A set of examples $E \subset \mathbf{X}$ can be considered to be the extension of a single table of a relational database. To simplify formal aspects, $\mathbf{X}$ is assumed to be finite in this work. All results are easily generalised to the case of continuous domains.

Examples are assumed to be sampled i.i.d with respect to a *distribution* $D : \mathbf{X} \to [0, 1]$. The probability to observe an instance $x \in \mathbf{X}$ under D is denoted as $P_{x \sim D}(x)$. The probability to observe an instance from a subset $W \subseteq \mathbf{X}$ is denoted as $P_D(W)$. If the underlying distribution is clear from the context we omit the subscripts.

Each example is assigned a label from the set $\mathbf{Y}$ of all possible labels by the *target function* $C : \mathbf{X} \to \mathbf{Y}$. We assume C to be fixed but unknown to the learner, whose task is to approximate it in a specified way. This work considers only supervised learning with a boolean target attribute $\mathbf{Y} = \{0, 1\}$.

2.2   Probabilistic Rules for Knowledge Representation

Encoding prior knowledge[1] is often done using any form of rules. For subgroup discovery Horn logic rules are the main representation language.

Definition 1. *A Horn logic rule consists of a body A, which is a conjunction of atoms over $A_1, \ldots, A_k$, and a head B, predicting a value for the target attribute. It is notated as $A \to B$. If the body evaluates to true the rule is said to be applicable, if the head evaluates to true, also, it is called correct.*

More generally a rule can be considered as a function $h : \mathbf{X} \to \mathbf{Y}$, assigning a prediction to each $x \in \mathbf{X}$.

For now we assume any form of prior knowledge to be represented by rules of this form. In subsection 4.3 we will see that the presented approach can easily be extended to incorporate any form of prior knowledge predicting the conditional distribution of the target variable.

Assuming $\mathbf{Y} = \{0, 1\}$, the following abbreviations are used:

$$h := \{x \in \mathbf{X} \mid h(x) = 1\}, \overline{h} := \mathbf{X} \setminus h$$
$$Y_+ := \{x \in \mathbf{X} \mid C(x) = 1\}, Y_- := \mathbf{X} \setminus Y_+$$

Using this notation, the Horn logic rules predicting a boolean target are of the form $(h \to Y_+)$ and $(h \to Y_-)$. Unlike for any strictly logical interpretation, rules are not expected to match the data exactly. Often it is sufficient if they point to regularities in the data. The intended semantics of a *probabilistic rule* is that the conditional probability $P(Y_+ \mid h)$ (or $P(Y_- \mid h)$) is higher than the class prior $P(Y_+)$ (or $P(Y_-)$). Probabilistic rules are often annotated by their corresponding conditional probabilities:

$$h \to Y_+ \ [0.8] \quad :\Leftrightarrow \quad P_{x \sim D}(C(x) = 1 \mid h(x) = 1) = 0.8$$

2.3   Performance Metrics

As a general task in supervised learning we want to estimate conditional probabilities of target attributes. Different performance metrics help to evaluate how useful and interesting single rules are. For the notion of interestingness different

---

[1] The term "prior knowledge" will be preferred to "background knowledge", because the latter is associated with precise knowledge for inference, while *prior knowledge* suggests a more probabilistic view.

formalisations have been proposed in the literature (e.g.[21]). In this work inter-estingness is considered equal to unexpectedness. This subsection collects some important metrics for rule selection.

The goal when training classifiers is to select a predictive model that separates positive and negative examples accurately.

Definition 2. *For a rule* $(A \rightarrow B)$ *the* accuracy *is defined as*

$$\mathbf{Acc}(A \rightarrow B) := P(A \cap B) + P(\overline{A} \cap \overline{B})$$

Definition 3. *The* precision *of a rule reflects the conditional probability that it is correct, given that it is applicable:*

$$\mathbf{Prec}(A \rightarrow B) := P(B \mid A)$$

Subgroup discovery focuses on rules covering subsets that – compared to the overall distribution – are biased in the data. The following metric has been used to measure interest in the domain of frequent itemset mining [3]. In the supervised context it measures the change in the target attribute's frequency for the subset covered by a rule.

Definition 4. *For any rule* $(A \rightarrow B)$ *the* **Lift** *is defined as*

$$\mathbf{Lift}(A \rightarrow B) := \frac{P(A \cap B)}{P(A)P(B)} = \frac{P(B \mid A)}{P(B)} = \frac{\mathbf{Prec}(A \rightarrow B)}{P(B)}$$

The **Lift** of a rule captures the value of "knowing" the prediction for estimating the probability of the target attribute. $\mathbf{Lift}(A \rightarrow B) = 1$ indicates that A and B are independent events. With $\mathbf{Lift}(A \rightarrow B) > 1$ the conditional probability of B given A increases, with $\mathbf{Lift}(A \rightarrow B) < 1$ it decreases.

During subgroup discovery rules are evaluated by a utility function. A pop-ular function is the following one, e.g. available in EXPLORA [10]:

Definition 5. *The* weighted relative accuracy *(***WRAcc***) of a rule* $(A \rightarrow B)$ *multiplies coverage* $P(A)$ *and bias* $P(B \mid A) - P(B)$*:*

$$\mathbf{WRAcc}(A \rightarrow B) := P(A) \cdot (P(B \mid A) - P(B))$$

The use of **WRAcc** as a measure for rule interestingness has been motivated elaborately in [13]. It is similar to the binomial test function, thus favours sig-nificant rules, but puts more emphasis on coverage [10]. Many other functions have been suggested in the literature [24,10], basically putting more emphasis on either coverage or bias.

3   Subgroup Discovery

Subgroup discovery aims at finding interesting subsets of the instance space that deviate from the overall distribution. The search is guided by a *utility function* that allows to find interesting rules with much smaller coverage and bias than possible with standard classifier induction algorithms. Subsection 3.1 briefly describes related work in subgroup discovery. How interesting rules interact, how to recognise redundant rules, and how to build single predictors from rulesets is discussed in 3.2. In subsection 3.3 incorporation of prior knowledge as a means to improve utility and diversity of the discovered rulesets is motivated. Subsection 3.4 shows a generic way of addressing subgroup discovery tasks using classifier induction algorithms.

3.1 Existing Approaches

The goal of subgroup discovery is to find interesting and novel patterns in datasets. Utility functions formalise a trade-off between the size of the subgroup and the unusualness in terms of a target attribute's observed frequency. There are two different strategies of searching for interesting rules: exhaustive and heuristic search.

MIDOS [24] and EXPLORA [10] tackle subgroup discovery by exhaustively evaluating the set of rule candidates. The set of rules are ordered by generality, which allows to prune large parts of the search space. The advantage of this strategy is that it allows to find the n best subgroups reliably. For the special case of exception rules similar exhaustive search strategies exists [22]. Finding subgroups on subsamples of the original data is a straightforward method to speed up the search process. As shown in [19,20] most of the utility functions commonly used for subgroup discovery are well suited to be combined with adaptive sampling. This sampling technique reads examples sequentially, continuously updating upper bounds for the sample errors, based on the data read so far. In this way, the required sample size allowing to give a probabilistic guarantee of not missing any of the n best subgroups can be reduced.

Heuristic search strategies are fast, but do not come with any guarantee to find the most interesting patterns. One recent example implementing a heuristic search is a variant of CN2. By adapting its evaluation measure for rule candidates to **WRAcc** the well known CN2 classifier has been turned into CN2-SD [12]. As a second modification the iterative cover approach of CN2 has been replaced by a heuristic weighting scheme. Example weights are either changed by a constant factor or by an additive term each time the example has been covered by a rule. In section 4 a new generic weighting scheme is proposed that allows to overcome some shortcomings of CN2-SD.

For pruning rulesets ROC analysis was suggested in [12]. According to the false positive and false negative rates all rules are plotted in ROC space [4]. Only rules lying on the convex hull are deemed relevant and may be turned into a single classifier by weighted majority vote. A major drawback of this filter is that it systematically discards one of two rules covering disjoint subsets and having almost the same performance. As soon as one of these rules is superior in both true positive and false negative rates, the other rule is considered to be redundant. This is not desirable in descriptive scenarios, as the only rule covering a specific subset of the instance space should not easily be discarded, nor for predictive settings, as diversity of base classifiers is crucial for reaching high predictive accuracy [2].

3.2   Combining Rules

There are different methods to combine a set of rules predicting the conditional probability of a target class. The approach put forward in this work is useful for descriptive and predictive settings, and it can be used to combine arbitrary predictors, especially rules represented in Horn logic. If the prediction of each rule is used to define a new attribute, then predictions can be combined by means of classifier induction techniques. The underlying assumption of Naïve Bayes [9] is that all attributes are conditionally independent given the class. These classifiers work surprisingly well in practice, often even if the underlying assumption is known to be violated. When mining rules iteratively, using the sampling technique proposed in section 4, the conditional independence assumption is not as unrealistic as one might expect. The reason is that all correlations "reported" by previously found patterns are "removed" from subsequently constructed samples.

Let $\{h_i : \mathbf{X} \to \mathbf{Y} \mid 1 \le i \le n\}$ denote a set of rules. Then for any given example $x \in \mathbf{X}$, labels $y_1, \ldots, y_n \in \mathbf{Y}$, and $h_1(x) = y_1, \ldots, h_n(x) = y_n$, the Naïve Bayes classifier estimates

$$
\begin{aligned}
&P(C(x) = y \mid h_1(x) = y_1, \ldots h_n(x) = y_n) \\
&= \frac{P(h_1(x) = y_1, \ldots, h_n(x) = y_n \mid C(x) = y) \cdot P(C(x) = y)}{P(h_1(x) = y_1, \ldots, h_n(x) = y_n)} \\
&\approx \frac{P(C(x) = y)}{P(h_1(x) = y_1, \ldots h_n(x) = y_n)} \prod_{1 \le i \le n} P(h_i(x) = y \mid C(x) = y) \\
&= \frac{P(C(x) = y) \prod_i P(h_i(x) = y_i)}{P(h_1(x) = y_1, \ldots, h_n(x) = y_n)} \prod_{1 \le i \le n} \frac{P(C(x) = y \mid h_i(x) = y)}{P(C(x) = y)} \\
&= \frac{P(C(x) = y) \prod_i P(h_i(x) = y_i)}{P(h_1(x) = y_1, \ldots, h_n(x) = y_n)} \prod_{1 \le i \le n} \mathbf{Lift}((h_i(x) = y_i) \to (C(x) = y))
\end{aligned}
$$

for each class $y \in \mathbf{Y}$. Especially for boolean $\mathbf{Y}$ it is easier to consider the ratios

$$
\begin{aligned}
\alpha(x) &:= \frac{P(Y_+ \mid h_1(x) = y_1, \ldots, h_n(x) = y_n)}{P(Y_- \mid h_1(x) = y_1, \ldots, h_n(x) = y_n)} \\
&= \frac{P(Y_+)}{P(Y_-)} \prod_{1 \le i \le n} \frac{\mathbf{Lift}((h_i(x) = y_i) \to Y_+)}{\mathbf{Lift}((h_i(x) = y_i) \to Y_-)},
\end{aligned}
\tag{1}
$$

as most of the terms cancel out, but we can still recalculate

$$
P(Y_+ \mid h_1(x) = y_1, \ldots, h_n(x) = y_n) = \frac{\alpha(x)}{1 + \alpha(x)}
$$

based on formula (1). So following the conditional independence assumption it is possible to combine rules to predict class probabilities, just knowing their **Lift** and the class priors. It is not necessary to restrict rules to the case in which the body evaluates to true. Please note that

$$\mathbf{Lift}(h \to Y_+) > 1 \;\Rightarrow\; \mathbf{Lift}(\overline{h} \to Y_-) > 1,$$

but the precisions of both rules may differ. So each rule $h \to Y_{+/-}$ should rather be considered to partition the instance space into h and $\overline{h}$, making a prediction for both subsets. As a consequence any two rules overlap. Thus, for any known degree of overlap between a rule R_1 that is part of the prior knowledge and a rule candidate R_2 under consideration, we have an expectation for $\mathbf{Lift}(R_1)$ based on $\mathbf{Lift}(R_2)$. This expectation reflects the assumption that R_2 does not introduce a **Lift** of its own, but simply shares a biased subset with R_1. If this assumption is met, then the rule candidate is redundant and should be ranked low. The **Lift** of each rule can be expressed relative to prior knowledge, e.g. of preceding rules. The following equation illustrates this idea for the simplified case of two rules and the subset $h_1 \cap h_2 \subset \mathbf{X}$:

$$\mathbf{Lift}((h_1, h_2) \to Y_+) = \frac{P(h_1, h_2 \mid Y_+)}{P(h_1, h_2)} = \frac{P(h_1 \mid Y_+) \cdot P(h_2 \mid h_1, Y_+)}{P(h_1) \cdot P(h_2 \mid h_1)}$$

$$= \mathbf{Lift}(h_1 \to Y_+) \cdot \underbrace{\frac{\mathbf{Lift}(h_2 \to (h_1, Y_+))}{\mathbf{Lift}(h_2 \to h_1)}}_{=:\mathbf{Lift}(h_2 \to Y_+ \mid h_1)}$$

The term $\mathbf{Lift}(h_2 \to Y_+ \mid h_1)$ can be regarded as the *relative* **Lift** of the rule $h_2 \to Y_+$ with respect to prior knowledge. It replaces $\mathbf{Lift}(h_2 \to Y_+)$ when estimating $\alpha(x)$ in formula (1) given $h_1 \to Y_+$. Applying the sampling technique introduced in section 4, rules with high relative performance are favoured. This usually results in rulesets with low redundancy and high diversity.

3.3   Iterative Subgroup Discovery

A drawback of classical subgroup discovery lies in a lack of expressiveness. Especially interesting *exceptions* to rules are hard to be detected using standard techniques, for mainly two reasons. First of all, due to the syntactical structure imposed by Horn logic it is often hard to exclude exceptions from rules, although this would improve the score assigned by the utility function. The syntactical bias is important, however, because we want the results to be understandable, and because it is the main reason for *diversity* within the n best subgroups. Without any syntactical restrictions the second best subgroup would usually be the best one after adding or removing a single example. The syntactical bias might not be sufficient to avoid sets of similar rules. Redundancy filters are a common technique to overcome this problem. Overlapping patterns like exceptions to rules are not found reliably that way. Exceptions could still be represented by separate rules. This fails for the second reason, namely that utility functions evaluate rules globally. Interactions between rules do not affect their scores.

Formalised prior knowledge like previously found patterns could help to refine existing utility measures. Two different approaches to exploit prior knowledge in the scope of subgroup discovery have been suggested so far. The first one is to prune rules violating a redundancy constraint [10]. This is possible during search, or as a post-processing step to present only the most interesting rules. With the ILP system RSD [11] another way of incorporating background knowledge has been proposed. It uses background knowledge to propositionalise relational data. For the learning step itself CN2-SD is used.

One of the advantages of the approach presented here is that it allows to turn any algorithm for training classifiers in the presence of noise into one for subgroup discovery with utility function **WRAcc** that can exploit prior knowledge. The next subsection shows a generic way to transform subgroup discovery tasks into classifier induction tasks, before a generic way to incorporate prior knowledge into supervised Data Mining is introduced in section 4.

3.4   Subgroup Discovery by Classifier Induction

This subsection briefly discusses the relation between subgroup discovery with utility function **WRAcc** and the task of classifier induction.

The goal of classifier induction is to select a predictive model that separates positive and negative examples with high predictive accuracy. Many algorithms and implementation exists for this purpose [16,23], basically differing in the set of models (hypothesis space H) and search strategies. Subgroup discovery is also a supervised learning task. Examples are classified with respect to a "property of interest". The overall goal is to find understandable and interesting rules, which is hard to be formalised. Thus, the process of model selection is guided by a utility function. In the following definition subgroup discovery is reduced to finding a single rule, only.

Definition 6. *Let H denote the set of models (rules) valid as output and D denote a distribution function over $\mathbf{X}$. The task of classifier induction is to find*

$$h^* := maxarg_{h \in H} \ \mathbf{Acc}(h).$$

For a given utility function $q : H \to \mathbb{R}$ the task of subgroup discovery is to find

$$h^* := maxarg_{h \in H} \ q(h).$$

For boolean target attributes common classifier induction algorithms do not benefit from finding rules with a precision below 50%. In contrast, for subgroup discovery it is sufficient if a class is observed with a frequency that is significantly higher than in the overall population. In cases of skewed class distributions the frequency in the covered subset might still be far below 50% for the most interesting rules. Choosing the utility function **WRAcc** we can transform subgroup discovery as defined above into classifier induction by a simple sampling technique to overcome imbalanced class distributions.

Definition 7. *For $D : \mathbf{X} \to [0, 1]$, $C : \mathbf{X} \to \mathbf{Y}$ let the* stratified random sample *distribution $D\prime$ of D (and C) be defined by*

$$P_{x \sim D\prime}(x) := \frac{P_{x \sim D}(x)}{|Y| \cdot P_{z \sim D}(C(z) = C(x))} = P_D(x) / \begin{cases} 2P_D(Y_+), & \text{for } C(x) = 1 \\ 2P_D(Y_-), & \text{for } C(x) = 0 \end{cases}$$

$D\prime$ is defined by rescaling D so that the class priors are equal.

Theorem 1. *For every rule $h \to Y_+$ the following equalities hold if $D\prime$ is the stratified random sample distribution of D:*

$$\mathbf{Acc}_{D\prime}(h \to Y_+) = 2\mathbf{WRAcc}_{D\prime}(h \to Y_+) - 1/2$$

$$= \mathbf{WRAcc}_D(h \to Y_+) \cdot \underbrace{\frac{1}{2P_D(Y_+) \cdot P_D(Y_-)}}_{\text{irrelevant for ranking rules}} - 1/2$$

Theorem 1 indicates that subgroup discovery tasks with utility function **WRAcc** can as well be solved by rule induction algorithms optimising predictive accuracy after a step of stratified resampling. A proof is given in the appendix. Further interesting relations between performance metrics are proven in [7].

4   Knowledge-Based Sampling

Before introducing techniques for sampling with respect to prior knowledge the task is formalised by a set of constraints.

4.1   Constraints for Resampling

After a first rough analysis has discovered global patterns we want to prepare a second iteration of Data Mining to find local patterns. The proposed idea is to construct samples that do not show the *biases* underlying previously discovered patterns, while taking care that all the remaining patterns remain intact.

Practically, for a given rule $R : h \to Y_+$ this means to consider a new distribution $D\prime$, as close to the original function D as possible. This is formalised by the following set of constraints. First of all, we want to remove the bias corresponding to R. In other words we want h and Y_+ to be independent:

$$P_{D\prime}(Y_+ \mid h) = P_{D\prime}(Y_+) \tag{2}$$

Next, we do not want the priors of h and Y_+ to change:

$$P_{D\prime}(h) = P_D(h) \tag{3}$$

$$P_{D\prime}(Y_+) = P_D(Y_+) \tag{4}$$

Finally, within each partition sharing the same class and prediction of R the new distribution is defined proportionally to the initial one:

$$P_{D'}(x \mid h \cap Y_+) = P_D(x \mid h \cap Y_+) \tag{5}$$
$$P_{D'}(x \mid h \cap Y_-) = P_D(x \mid h \cap Y_-) \tag{6}$$
$$P_{D'}(x \mid \overline{h} \cap Y_+) = P_D(x \mid \overline{h} \cap Y_+) \tag{7}$$
$$P_{D'}(x \mid \overline{h} \cap Y_-) = P_D(x \mid \overline{h} \cap Y_-) \tag{8}$$

Given a database and a global pattern R we can apply any Data Mining technique after sampling with respect to D'. This might ease the detection of further patterns. An advantage of mining the resampled data rather than a dataset without the covered examples shows, if there are further patterns within the covered subset. These patterns can still be observed after resampling, just rescaled proportionally. This helps to find exceptions to successful rules, as motivated in subsection 3.3, or patterns overlapping in some other way.

Please note, that a subgroup pattern showing in the new sample may be interesting relative to the prior knowledge, only. Let

$$P(Y_+ \mid A) = P(Y_+) = 0.5 \quad \text{for a rule} \quad A \to Y_+.$$

$\mathbf{Y}$ is distributed in A just as in the overall population, so this rule would not be deemed interesting by any reasonable utility function. Now assume that in the prior knowledge there is a statement about a superset of A:

$$B \to Y_+ \ [0.9] \quad \text{with} \quad A \subset B.$$

This rule predicts a higher conditional probability of Y_+ given B. In this context the rule $(A \to Y_+)$ becomes interesting as an exception to the prior knowledge, because we would rather expect $P(Y_+ \mid A) = P(Y_+ \mid B)$. The reason is that the prediction for $B \subset X$ is more specific than the general class priors. In general switching from the initial distribution to the resampled data is a step of applying prior knowledge by means of sampling. This step allows to find overlapping and nested patterns sequentially.

4.2   Constructing a New Distribution Function

In subsection 4.1 the idea of sampling with respect to an altered distribution function has been presented. Intuitively, prior knowledge and known patterns are "filtered out". This subsection proves that the proposed constraints (2) to (8) induce a unique target distribution.

Definition 8. *The lift of an example $x \in \mathbf{X}$ for a rule $(h \to Y_+)$ is defined as*

$$\mathbf{Lift}(h \to Y_+, x) := \begin{cases} \mathbf{Lift}(h \to Y_+), & \text{for } x \in h \cap Y_+ \\ \mathbf{Lift}(h \to Y_-), & \text{for } x \in h \cap Y_- \\ \mathbf{Lift}(\overline{h} \to Y_+), & \text{for } x \in \overline{h} \cap Y_+ \\ \mathbf{Lift}(\overline{h} \to Y_-), & \text{for } x \in \overline{h} \cap Y_- \end{cases}$$

Theorem 2. *For any initial distribution D and given rule R the probability distribution $D\prime$ is induced uniquely by the constraints (2) to (8) as follows:*

$$P_{D\prime}(x) := P_D(x) \cdot (\mathbf{Lift}_D(R, x))^{-1}$$

Proof. The proof is exemplarily shown for the partition $(h \cap Y_+)$, in which the rule under consideration is both applicable and correct. $D\prime$ can be rewritten in terms of D and $\mathbf{Lift}(R, x)$, assuming that the constraints hold:

$$
\begin{aligned}
(\forall x \in h \cap Y_+) : P_{D\prime}(x) &= P_{D\prime}(x \mid h \cap Y_+) \cdot P_{D\prime}(h \cap Y_+) \\
&= P_D(x \mid h \cap Y_+) \cdot P_{D\prime}(h) \cdot P_{D\prime}(Y_+) \\
&= \frac{P_D(x)}{P_D(h \cap Y_+)} \cdot P_D(h) \cdot P_D(Y_+) \\
&= P_D(x) \cdot (\mathbf{Lift}_D(h \to Y_+))^{-1}
\end{aligned}
$$

The other three partitions can be rewritten analogously. On the other hand, it can easily be validated that $D\prime$ as defined by theorem 2 is in fact a distribution satisfying constraints (2) to (8):

$$P_{D\prime}(h \cap Y_+) = P_D(h \cap Y_+) \cdot (\mathbf{Lift}_D(R, x))^{-1} = P_D(h) \cdot P_D(Y_+)$$

and analogously for the other partitions. This directly implies constraints (2) to (4) by marginalising out. Constraints (5) to (8) are met, because for all four partitions $D\prime$ is defined proportionally to D. This implies that the conditional probabilities given the partitions are equivalent.

4.3   Weighting Examples Using Prior Knowledge

In the last subsection it was discussed how to alter an initial distribution in the presence of prior knowledge. The goal is to construct samples not reflecting previously found patterns anymore. This idea stems from boosting classifiers, which was also first introduced in terms of altering an initial distribution function and a corresponding sampling technique [17]. The idea of boosting is to repeatedly apply a "weak" base learner and to combine the predictions. The probabilities of examples are adjusted in such a way that in later iterations the weak learner has to focus on the "hard" examples not yet covered sufficiently by the ensemble of base classifiers.

As a general alternative to resampling it is possible to assign weights to examples, reflecting a change in the underlying distribution. This method is common in boosting literature to avoid resampling [5,6,18]. It can be understood in terms of importance sampling [14]: The example set is assumed to be drawn independently from an initial distribution D. Then each example x is assigned the weight $D\prime(x)/D(x)$ rather than sampling directly with respect to $D\prime$, which may be infeasible.

For subgroup discovery the use of weighted examples may be less appropriate, as even uniformly distributed subsets may be represented as a single example with high weight. On the other hand, for given example weights resampling can easily be performed by a Monte Carlo technique called rejection sampling [14]. A straight-forward implementation of this technique has successfully been applied to cost-sensitive learning [25], which is very similar from a technical point of view. In this subsection a knowledge-based weighting scheme is introduced. It can replace resampling if all subsequently applied algorithms are capable of using example weights, and if it meets the requirements of the learning task. In other cases it can still be used as a basis for rejection sampling.

Theorem 2 defines a new distribution to sample from, given a single rule R as prior knowledge. The following strategy for weighting examples is more general. First of all the number of classes $|\mathbf{Y}|$ is not restricted to two. As a second generalisation the prior knowledge θ may be of arbitrary form. It is assumed to be associated to a function

$$\hat{P}(x, y, \theta) = \hat{P}(C(x) = y \mid x, \theta) \approx P(C(x) = y \mid x)$$

estimating probabilities for each $\langle x, y \rangle \in \mathbf{X} \times \mathbf{Y}$. Assuming the class priors $P(C(x) = y)$ to be known for each $y \in \mathbf{Y}$ and applying the definition of the **Lift** the corresponding *estimated* **Lift** can easily be computed as

$$\widehat{\mathbf{Lift}}(x, \theta) := \frac{\hat{P}(x, C(x), \theta)}{P_{z \sim D}(C(z) = C(x))}$$

Given a procedure for sampling examples $x \sim D$ independently, the following distribution generalising theorem 2 can be used for weighting each example:

$$P_{D\prime}(x) := P_D(x) \cdot (\widehat{\mathbf{Lift}}(x, \theta))^{-1} \tag{9}$$

To remove prior probabilistic knowledge from a data stream applying formula (9) it is sufficient to assign each example x from the stream a weight of $\widehat{\mathbf{Lift}}(x, \theta)^{-1}$, as the factor $D(x)$ is already accounted for by sampling with respect to D.

5   Experiments

The proposed idea of subgroup discovery utilising all forms of previously discovered patterns has been evaluated on three datasets from the UCI Machine Learning Library [1] and a sample of the KDD Cup 2004 Quantum Physics dataset[2]. For simplicity attributes with missing values have been discarded. All datasets have boolean target attributes. Further characteristics are listed in table 1.

Three subgroup discovery algorithms have been integrated into the learning environment YALE [15]. For mining subgroup rules from samples the embedded WEKA [23] rule induction algorithm has been applied to stratified samples,

---

[2] http://kodiak.cs.cornell.edu/kddcup/

Dataset	Examples	# Nominal Attr.	# Numerical Attr.	Minority class
Quantum Physics	10.000	–	71	50.0%
Ionosphere	351	–	34	35.8%
Credit Domain	690	6	9	44.5%
Mushrooms	8.124	22	–	48.2%

Table 1. Datasets used for experimental evaluation.

which is valid due to theorem 1. The algorithm CONJUNCTIVERULE heuristically selects a single Horn logic rule with high predictive accuracy, which translates into high **WRAcc**. It is applied repeatedly by the subgroup discovery algorithms. The *knowledge-based sampling algorithm* (KBS) applies sampling as presented in section 4. Rules are combined as discussed in subsection 3.2, similar to the Naïve Bayes method. KBS is compared to two other reweighting strategies reported in the subgroup discovery literature [11]. After a positive example e has been covered by i rules its new weight is computed as

$$w_i(e) := \frac{1}{i+1} \text{ (additive), or } w_i(e) := \gamma^i \text{ for given } \gamma \in [0,1] \text{ (multiplicative).}$$

Accordingly, two versions of subgroup discovery ruleset induction (SDRI) have been implemented, which are similar to CN2-SD. The variant that applies CONJUNCTIVERULE on stratified samples after *additive* reweighting is referred to as SDRI$^+$, the one with *multiplicative* reweighting as SDRI*. Reweighting is performed iteratively. The class explicitly predicted by a rule is defined to be the positive one, as fixing one of the classes as positive gave worse experimental results. The rulesets constructed by SDRI are combined to a single probabilistic prediction as in CN2-SD: The predicted target class distributions of all applicable rules are averaged.

The goal of subgroup discovery in this setting is to find a small set of (understandable) rules, giving a good picture of the data. In more formal terms the probabilistic classifiers built from the rulesets should be accurate. This property is measured by the area under the ROC curve metric (AUC) [4].

Figure 1 to 4 show how the AUC metric changes with an increasing number of rules. All values have been estimated by 10fold cross-validation. The default for the parameter γ of SDRI* was set to 0.9 as suggested in [11]. For all but the mushrooms dataset this value gave best results[3]. For mushrooms the results for the better value $\gamma = 0.7$ are reported. For a higher value of γ it generally took more iterations to reach a similar AUC performance, for lower values the algorithm converged more quickly, but reached worse results.

In all figures the KBS algorithm outperforms SDRI with both reweighting strategies, while none of the SDRI variants is clearly superior to the other one. In figure 1 all three algorithms manage to find useful rules repeatedly. SDRI$^+$ performs best for sets of 3 to 6 rules. For larger rulesets KBS is superior. SDRI* performs worst. Figure 2 shows the performance for a smaller

---

[3] The parameter was empirically decreased in steps of 0.1 and increased to 0.95.

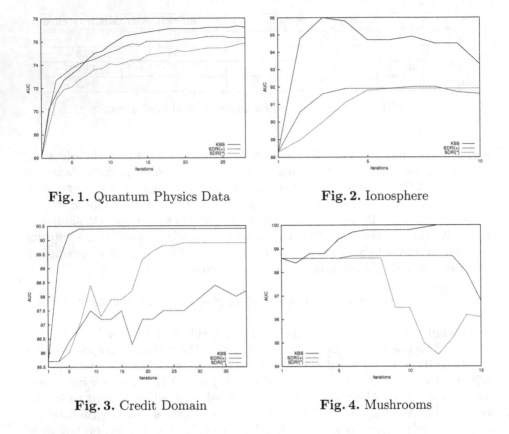

Fig. 1. Quantum Physics Data   **Fig. 2.** Ionosphere

Fig. 3. Credit Domain   **Fig. 4.** Mushrooms

dataset. Again KBS performs best, although it overfits after the 3rd iteration. The SDRI variants reach their maxima later. This delay is even more significant for the credit domain data, illustrated in figure 3. After iteration 7 the predictions of KBS remain constant, while the AUC values of the SDRI rulesets improve non-monotonically and are still significantly worse after 40 iterations. Finally, in the experiment shown in figure 4 KBS reaches 100% AUC with just 12 rules, while SDRI does not manage to improve over the performance of the first rule at all. As a further experiment on this dataset ADABOOST has been run on top of CONJUNCTIVERULE. After 15 iterations it still has an error rate of about 2.5%.

Table 2 lists the average performance of rulesets. For the ionosphere and credit domain dataset the number of rules with best performance regarding AUC was chosen. For the KDD Cup data (Quantum Physics) the number of rules was set to 15. The ROC filter for rulesets discussed in subsection 3.2 was applied to both SDRI variants, denoted as RF in table 2. As mentioned in subsection 4.1 some patterns are interesting relative to prior knowledge, only. The columns **AvgCov** and **AvgWRAcc** in table 2 demonstrate that absolute values of performance metrics may be misleading regarding how well rules are suited to predict a target class. **AvgCov** denotes the average coverage of rules,

Dataset	Algorithm	# Rules	AUC	AvgCov	AvgWRAcc
Ionosphere	KBS	3	96.0 ($\pm$ 3.0)	42.7%	0.121
Ionosphere	SDRI$^+$	7	92.0 ($\pm$ 7.4)	37.6%	0.120
Ionosphere	SDRI$^+$, RF	4	91.7 ($\pm$ 7.0)	35.3%	0.120
Ionosphere	SDRI*	6	91.9 ($\pm$ 7.3)	60.1%	0.123
Ionosphere	SDRI*, RF	3	91.0 ($\pm$ 6.7)	40.6%	0.119
Credit Domain	KBS	7	90.4 ($\pm$ 3.4)	42.2%	0.057
Credit Domain	SDRI$^+$	31	88.4 ($\pm$ 4.2)	56.8%	0.156
Credit Domain	SDRI$^+$, RF	3	87.0 ($\pm$ 5.3)	66.9%	0.139
Credit Domain	SDRI*	27	89.9 ($\pm$ 4.0)	55.8%	0.164
Credit Domain	SDRI*, RF	2	85.7 ($\pm$ 5.3)	66.9%	0.139
Quantum Physics	KBS	15	76.8 ($\pm$ 1.2)	38.6%	0.023
Quantum Physics	SDRI$^+$	15	76.0 ($\pm$ 1.9)	50.5%	0.054
Quantum Physics	SDRI$^+$, RF	12	74.3 ($\pm$ 2.0)	50.0%	0.056
Quantum Physics	SDRI*	15	74.8 ($\pm$ 2.1)	42.7%	0.071
Quantum Physics	SDRI*, RF	8	74.2 ($\pm$ 2.1)	44.7%	0.074

Table 2. Performance values for different subgroup algorithms.

AvgWRAcc the average weighted relative accuracy. Global evaluation rewards overlapping rules for reporting the same pattern multiple times, while rules capturing smaller patterns not covered by any other rule may perform worse if evaluated stand-alone. This explains why both the average absolute coverage and absolute **WRAcc** of KBS is lower for two of the three datasets than the corresponding values of SDRI, but the AUC values are still higher. The ROC filter generally seems to neither improve the AUC score nor the average global utility function. In most cases it prunes the ruleset at the price of a reduced performance. Increasing coverage is comparably trivial.

As an overall result the experiments show that knowledge-based sampling helps to shift the focus of subgroup discovery to yet undiscovered patterns, allowing to find a small number of rules that help to build accurate probabilistic classifiers. Rulesets with higher average values of utility functions that were not constructed to maximise diversity turn out to be less accurate.

6   Conclusion

In this work local pattern mining was defined in terms of prior knowledge available to a learner. Subgroup discovery was identified as a matching learning task, but the available algorithms do not incorporate previously discovered patterns and prior domain knowledge into their utility functions. In section 4 a generic way of incorporating prior knowledge by means of sampling was presented. The selected samples do no longer reflect the prior knowledge and can be used to mine further local patterns. Applying the utility function to such a sample means not to reward rules for overlapping with previously known biased subsets, but to rank rules by their new own contribution. This helps to focus on rulesets that are almost orthogonal, thus the conditional independence assumption is not as unrealistic as in general. As a consequence, rules predicting the conditional probabilities of a target attribute can well be combined by the Naïve Bayes strategy. The simplicity of the reweighting scheme allows to interpret the found patterns either globally or in their specific context, based on the intuitive **Lift** measure. To simplify subgroup discovery it was shown how to address pattern mining with utility function **WRAcc** with common rule induction algorithms. The developed subgroup discovery algorithm has been validated experimentally and shown to outperform existing reweighting and rule combination strategies in the scope of subgroup discovery.

References

1. C.L. Blake and C.J. Merz. UCI repository of machine learning databases, 1998. http://www.ics.uci.edu/~mlearn/MLRepository.html.
2. Leo Breiman. Random forests. *Machine Learning*, 45(1):5–32, 2001.
3. Sergey Brin, Rajeev Motwani, Jeffrey Ullman, and Shalom Tsur. Dynamic Itemset Counting and Implication Rules for Market Basket Data. In *Proceedings of ACM SIGMOD Conference on Management of Data (SIGMOD '97)*, pages 255–264, Tucson, AZ., 1997. ACM.
4. T. Fawcett. ROC Graphs: Notes and Practical Considerations for Researchers, 2004. Submitted to Machine Learning.
5. Yoav Freund and Robert R. Schapire. A decision–theoretic generalization of on-line learning and an application to boosting. *Journal of Computer and System Sciences*, 55(1):119 – 139, 1997.
6. J. H. Friedman, T. Hastie, and R. Tibshirani. Additive logistic regression: A statistical view of boosting. *Annals of Statistics*, (28):337–374, 2000.
7. Johannes Fürnkranz and Peter A. Flach. An Analysis of Rule Evaluation Metrics. In *Proceedings of the 20th International Conference on Machine Learning (ICML-03)*. Morgen Kaufman, 2003.
8. David Hand. Pattern detection and discovery. In David Hand, Niall Adams, and Richard Bolton, editors, *Pattern Detection and Discovery*. Springer, 2002.
9. George H. John and Pat Langley. Estimating continuous distributions in Bayesian classifiers. In *Proceedings of the Eleventh Conference on Uncertainty in Artificial Intelligence*, pages 338–345. Morgan Kaufmann, 1995.
10. Willi Klösgen. Explora: A Multipattern and Multistrategy Discovery Assistant. In Usama M. Fayyad, Gregory Piatetsky-Shapiro, Padhraic Smyth, and Ramasamy Uthurusamy, editors, *Advances in Knowledge Discovery and Data Mining*, chapter 3, pages 249–272. AAAI Press/The MIT Press, Menlo Park, California, 1996.
11. N. Lavrac, F. Zelezny, and P. Flach. RSD: Relational subgroup discovery through first-order feature construction. In *12th International Conference on Inductive Logic Programming*. Springer, 2002.
12. Nada Lavrac, Peter Flach, Branko Kavsek, and Ljupco Todorovski. Rule Induction for Subgroup Discovery with CN2-SD. In Marko Bohanec, Dunja Mladenic, and Nada Lavrac, editors, *2nd Int. Workshop on Integration and Collaboration Aspects of Data Mining, Decision Support and MetaLearning*, August 2002.
13. Nada Lavrac, Peter Flach, and Blaz Zupan. Rule Evaluation Measures: A Unifying View. In *9th International Workshop on Inductive Logic Programming*, Lecture Notes in Computer Science. Springer, 1999.

14. D.J.C. Mackay. Introduction To Monte Carlo Methods. In *Learning in Graphical Models*, pages 175–204. 1998.
15. Ingo Mierswa, Ralf Klinkberg, Simon Fischer, and Oliver Ritthoff. A Flexible Platform for Knowledge Discovery Experiments: YALE – Yet Another Learning Environment. In *LLWA 03 - Tagungsband der GI-Workshop-Woche Lernen - Lehren - Wissen - Adaptivität*, 2003.
16. Tom M. Mitchell. *Machine Learning*. McGraw Hill, New York, 1997.
17. Robert E. Schapire. The Strength of Weak Learnability. *Machine Learning*, 5:197–227, 1990.
18. Robert E. Schapire and Yoram Singer. Improved boosting using confidence-rated predictions. *Machine Learning*, 37(3):297–336, 1999.
19. Tobias Scheffer and Stefan Wrobel. A Sequential Sampling Algorithm for a General Class of Utility Criteria. In *Proceedings of the International Conference on Knowledge Discovery and Data Mining*, 2000.
20. Tobias Scheffer and Stefan Wrobel. Finding the Most Interesting Patterns in a Database Quickly by Using Sequential Sampling. *Journal of Machine Learning Research*, 3:833–862, 2002.
21. Avi Silberschatz and Alexander Tuzhilin. What makes patterns interesting in knowledge discovery systems. *IEEE Transactions on Knowledge and Data Engineering*, 8(6):970–974, dec 1996.
22. Einoshin Suzuki. Discovering Interesting Exception Rules with Rule Pair. In *ECML/PKDD 2004 Workshop, Advances in Inductive Rule Learning*, 2004.
23. Ian Witten and Eibe Frank. *Data Mining – Practical Machine Learning Tools and Techniques with Java Implementations*. Morgan Kaufmann, 2000.
24. Stefan Wrobel. An Algorithm for Multi–relational Discovery of Subgroups. In J. Komorowski and J. Zytkow, editors, *Principles of Data Mining and Knowledge Discovery: First European Symposium (PKDD 97)*, pages 78–87, Berlin, New York, 1997. Springer.
25. Bianca Zadrozny, John Langford, and Abe Naoki. Cost–Sensitive Learning by Cost–Proportionate Example Weighting. In *Proceedings of the 2003 IEEE International Conference on Data Mining (ICDM'03)*, 2003.

Appendix

We repeat the definition of the two tasks, substituting C for Y_+ (or Y_-) and $\overline{C}$ for Y_- (or Y_+). H denotes a set of valid Horn logic rules with head C.

Classification Find an $h \in H$ maximising *predictive accuracy*:

$$\mathbf{Acc}(h \rightarrow C) = P(h \cap C) + P(\overline{h} \cap \overline{C})$$

Subgroup Discovery with WRAcc Find an $h \in H$ maximising

$$\mathbf{WRAcc}(h \rightarrow C) = P(h) \cdot (P(C|h) - P(C))$$

The correctness of the theorem is shown using two lemmas.

Lemma 1. *The two tasks are equivalent, if and only if the priors of both class labels are equal:*

$$P(C) = P(\overline{C}) = 1/2$$

Proof. First we rewrite predictive accuracy:

$$Acc(h, C) = P(h \cap C) + P(\overline{h} \cap \overline{C}) = P(h \cap C) + \left(P(\overline{h}) - P(\overline{h} \cap C)\right)$$
$$= P(h \cap C) + P(\overline{h}) - (P(C) - P(h \cap C)) = 2P(h \cap C) + P(\overline{h}) - P(C)$$
$$= 2P(C|h)P(h) + 1 - P(h) - P(C) = 2P(h)\left(P(C|h) - 1/2\right) - P(\overline{C}) \quad (10)$$

The order of rules according to this metric does not change if we drop the constant additive terms $P(\overline{C})$ and the constant factor of 2 in formula (10), so

$$argmax_{h \in H} Acc(h, C) = argmax_{h \in H} \left(P(h) \cdot (P(C|h) - 1/2)\right)$$

Obviously this is equivalent to **WRAcc** if and only if $P(C) = 1/2$. In this case the corresponding rankings of H are equivalent.

If the condition of lemma 1 is violated for the original distribution D we can perform stratified sampling using definition 7:

$$P_{D\prime}(x) := \frac{P_D(x)}{2P_D(C(x))} \quad (11)$$

Considering a sample from $D\prime$ as defined by (11) we expect $P_{D\prime}(h)$ and $P_{D\prime}(C|h)$ to differ from $P_D(h)$ and $P_D(C|h)$, respectively. As the following lemma states such samples are nevertheless appropriate for rule selection.

Lemma 2. *The order of a ruleset H induced by the WRAcc metric is equivalent for any two distributions D and $D\prime$, as long as formula (11) holds.*

Proof. Let us first rewrite $P_{D\prime}(h)$ in terms of D:

$$P_{D\prime}(h) = \frac{P_D(h \cap C)}{2P_D(C)} + \frac{P_D(h \cap \overline{C})}{2P_D(\overline{C})} = \frac{P_D(h)}{2}\left(\frac{P_D(h \cap C)}{P_D(h)P_D(C)} + \frac{P_D(h \cap \overline{C})}{P_D(h)P_D(\overline{C})}\right)$$
$$= P_D(h) \cdot \underbrace{\frac{1}{2}\left(\mathbf{Lift}_D(h \to C) + \mathbf{Lift}_D(h \to \overline{C})\right)}_{=:\alpha} \quad (12)$$

Having $P_{D\prime}(h) = P_D(h) \cdot \alpha$ allows to reformulate **WRAcc**$_{D\prime}$ like this:

$$\mathbf{WRAcc}_{D\prime}(h \to C) = P_{D\prime}(h) \cdot (P_{D\prime}(C|h) - P_{D\prime}(C))$$
$$= P_{D\prime}(h) \cdot \left(\frac{P_{D\prime}(C \cap h)}{P_{D\prime}(h)} - 1/2\right) = P_D(h) \cdot \alpha \cdot \left(\frac{\frac{P_D(C \cap h)}{2P_D(C)}}{P_D(h) \cdot \alpha} - 1/2\right)$$
$$= P_D(h) \cdot \alpha \cdot \left(\frac{1}{2}\frac{P_D(C \cap h)}{P_D(C) \cdot P_D(h) \cdot \alpha} - 1/2\right)$$
$$= \frac{1}{2}P_D(h)\left(\mathbf{Lift}_D(h \to C) - \alpha\right) \quad (13)$$

Formula (13) can be simplified by rewriting α, exploiting that

$$\mathbf{Lift}_D(h \to \overline{C}) = \frac{1 - P_D(C|h)}{P_D(\overline{C})} = \frac{1}{P_D(\overline{C})} - \frac{P_D(C)}{P_D(\overline{C})} \cdot \mathbf{Lift}_D(h \to C) \quad (14)$$

After plugging (14) into α we receive

$$\alpha = 1/2 \cdot \left(\mathbf{Lift}_D(h \to C) + \frac{1}{P_D(\overline{C})} - \frac{P_D(C)}{P_D(\overline{C})} \cdot \mathbf{Lift}_D(h \to C) \right)$$

$$= 1/2 \cdot \left(\left(1 - \frac{P_D(C)}{P_D(\overline{C})} \right) \mathbf{Lift}_D(h \to C) + \frac{1}{P_D(\overline{C})} \right)$$

$$= \frac{1}{2P_D(\overline{C})} \cdot \left((P_D(\overline{C}) - P_D(C)) \, \mathbf{Lift}_D(h \to C) + 1 \right)$$

$$= \frac{1}{2P_D(\overline{C})} \cdot \left((1 - 2P_D(C)) \, \mathbf{Lift}_D(h \to C) + 1 \right)$$

which can now be substituted into (13):

$$\frac{1}{2} P_D(h) \cdot (\mathbf{Lift}_D(h \to C) - \alpha)$$

$$= \frac{1}{2} P_D(h) \cdot \left(\mathbf{Lift}_D(h \to C) - \frac{(1 - 2P_D(C)) \, \mathbf{Lift}_D(h \to C) + 1}{2P_D(\overline{C})} \right)$$

$$= \frac{1}{2} P_D(h) \cdot \left(\mathbf{Lift}_D(h \to C) \left(1 - \frac{1 - 2P_D(C)}{2 - 2P_D(C)} \right) - \frac{1}{2P_D(\overline{C})} \right)$$

$$= \frac{1}{2} P_D(h) \cdot \left(\mathbf{Lift}_D(h \to C) \frac{1}{2 - 2P_D(C)} - \frac{1}{2P_D(\overline{C})} \right)$$

$$= \frac{1}{4P_D(\overline{C})} \cdot P_D(h) \cdot (\mathbf{Lift}_D(h \to C) - 1)$$

$$= \frac{1}{4P_D(\overline{C}) \cdot P_D(C)} \cdot P_D(h) \cdot (P_D(C|h) - P_D(C))$$

$$= \underbrace{\frac{1}{4P_D(\overline{C}) \cdot P_D(C)}}_{\text{irrelevant}} \cdot \mathbf{WRAcc}_D(h \to C) \tag{15}$$

The constant factor on the left hand side does not change the ranking of rulesets. We may drop it and end up with the definition of the **WRAcc** metric for D, which completes the proof of lemma 2.

Putting together formulas (15) and (10) we receive

$$\mathbf{Acc}_{D\prime}(h \to C) = 2P_{D\prime}(h) \left(P_{D\prime}(C|h) - 1/2 \right) - P_{D\prime}(\overline{C})$$

$$= 2P_{D\prime}(h) \left(P_{D\prime}(C|h) - P_{D\prime}(C) \right) - 1/2 = 2\mathbf{WRAcc}_{D\prime}(h \to C) - 1/2$$

$$= \frac{1}{2P_D(\overline{C}) \cdot P_D(C)} \cdot \mathbf{WRAcc}_D(h \to C) - 1/2,$$

which proves theorem 1.

Temporal Evolution and Local Patterns

Myra Spiliopoulou and Steffan Baron*

Otto-von-Guericke-Universität Magdeburg
{sbaron,myra}@iti.cs.uni-magdeburg.de

Abstract. We elaborate on the subject of pattern change as a result of population evolution. We provide an overview of literature threads relevant to this subject, where the focus is on related works in the area of pattern adaptation rather than on modelling or understanding change. We then describe our temporal model for patterns as evolving objects and propose criteria to capture the interestingness of pattern change. We also present heuristics that trace interesting changes.

1 Introduction

For several years, data mining has concentrated on the discovery of patterns upon a stationary population. However, most populations under observation in knowledge discovery are more often than not, subject to changes. Some of those changes are due to factors internal to the population, such as aging, while others are caused by external forces. Actions emanating from the results of knowledge discovery also cause changes on population behaviour. Hence, pattern changes due to external or internal influences should become integral part of the derived knowledge on a population.

In this study we elaborate on *global* and *local* patterns over a population evolving across the time axis. As Morik and Köpke point out in [30], local patterns are not yet clearly defined. They term patterns that describe rare events and deviate from a global model as local. Similarly, Pensa and Boulicaut put emphasis on the fact that local patterns describe only part of the database as they do not provide a "global picture" [38]. In this paper, we use a combination of the aforementioned definitions: local patterns cover small parts of the data space and deviate from the distribution of the population as a whole. In our context, the data space incorporates the temporal dimension. Hence, a local pattern characterises parts of the population for only limited time periods. Respectively, global patterns are present during the whole lifetime of the population. Notwithstanding this notion of locality, both global and local patterns show variations and may exhibit trends.

When patterns are observed as temporal objects describing an evolving population, their conventional properties, over a stationary data-space, must be reconsidered. Questions include: What properties uniquely identify a pattern?

---

* Work of this author has been performed during his PhD at the Humboldt-Universität zu Berlin.

K. Morik et al. (Eds.): Local Pattern Detection, LNAI 3539, pp. 190–206, 2005.

When does a pattern cease to exist and when can we say that it has only under-gone change? How do we distinguish between a local pattern and noise in the population?

In this work, we build upon our previous results on modelling patterns as temporal objects with variant and invariant properties across the time axis. From the above questions, we concentrate on capturing and modelling *pattern change* and use this model as basis for the properties of pattern locality and stability. To distinguish between local patterns and noise, we use heuristics that assess the interestingness of pattern changes. We focus on association rules and frequent sequences, i.e. on patterns rather than models. We do not assume or seek a global model for any time period.

In the first part of this paper, we discuss research threads on pattern adap-tation over a changing population and on pattern interestingness. In the second part, starting with section 3, we present our own approach, encompassing a tem-poral model and a framework to describe and assess pattern interestingness. We use our model to analyse a mail server log and report our findings. The last section concludes the study.[1]

2   Research Threads on the Subject of Pattern Change

Research associated with pattern change comes from different threads. Concept and population drift is investigated in the context of adaptive methods, in which patterns are aligned as new data is accumulated. Incremental mining methods serve the same objective of pattern adaptation in the field of unsupervised learn-ing, although adhering to different priorities and paying particular emphasis to algorithmic efficiency. In principle, such methods are suitable for the discovery of local patterns. However, in contrast to our approach they consider all data collected so far when updating patterns. The drawback of this approach lays in the constantly growing time lag between data change and the detection of its effects on the patterns.[2] Closely related to our work is the research on the detec-tion of differences among datasets. Pattern changes occur across the time axis, so temporal mining results are of obvious relevance here. Finally, a number of studies on the interestingness of patterns are devoted to discovering interesting pattern changes.

A survey of all relevant literature would go beyond the scope of this study. Rather we discuss a selection of related results, organised into three major cate-gories: a category on "knowledge alignment" which encompasses methods for the adaptation of patterns to change; the "change detection" category that includes indicators and supportive algorithms to discover pattern changes; and, finally, a category on "detection of interesting changes" that contains citations on in-terestingness in general but focuses more on interestingness for pattern changes,

---

[1] This paper contains and extends results reported in [7,8] and [4].
[2] Due to space constraints a detailed discussion of these methods is not included here but can be found in [6].

temporal patterns and temporal associations between patterns, including the validity of temporal rules.

2.1 Knowledge Alignment

The phenomenon of pattern change in populations has led to methods for knowledge alignment under the label "adaptive methods", mostly for classification.

Concept drift has been the subject of intensive research, both in its theoretical underpinnings and in the context of specific applications, e.g. security or user preferences and customer retention. Much research on concept drift comes from the domain of supervised learning. Classification upon data streams, i.e. sequences of records, is studied in [20,41], whereby Hulten et al. gradually generate a new classifier as the old one becomes inaccurate [20], while Street and Kim employ an ensemble of classifiers that consider all data points [41]. The temporal aspect of concept drift is stressed by Koychev in [24].

There are various paradigms for classification. The methods of [20,41] are based on decision trees. Support vector machines are considered in [22,43], while Morik and Rüping focus on inductive logic programming [31]. Inductive inference upon decision trees is proposed in [19]. The approach of Case et al. is designed for different types of learners [10].

An interesting type of knowledge alignment concerns user profiles in information retrieval. For example, in [36,47]: It is recognised that user preferences change, both because document corpora drift towards new foci and because the interests of the person themselves are not static. The algorithm "Alipes" by [47] stresses the difference between long-term and short-term profiles, the latter corresponding roughly to local patterns across the time axis. Both works use user feedback to align the profiles: Alipes recognises whether the change refers to the long-term or the short-term profile and adapts accordingly or even replaces an existing short-term profile with a new one [47]. Pazzani and Billsus acquire user feedback with respect to a predefined set of topics and then adapt a classifier to it [36], similarly to the aforementioned methods.

In [14], Wei Fan addresses the problem of selecting data chunks to mine a concept-drifting data stream. The motivation lays in the fact that old data may or may not be an appropriate basis for the classification of the most recent data in the stream. The proposed algorithm uses information gain to select the appropriate features from each data chunk and builds a classifier ensemble that considers multiple data chunks and multiple features.

The goal of knowledge alignment is to reflect the properties of a changing population. Perfect adaptation can be achieved by aligning the patterns to each new record. However, this incurs a runtime overhead that is intolerable and potentially unnecessary for some applications. Moreover, in the identification and interpretation of change, the distinction between stationary properties of the population and local patterns is not possible.

2.2   Change Detection

Orthogonally to adaptive methods, algorithms and indicators have been proposed, most of them in the field of unsupervised learning, which intend to detect population change. These contributions can roughly be subdivided based on the schema described in [23]. They identify three types of change indicators; performance measures, properties of the model and properties of the data. The first type is specific to supervised learning. It corresponds to measures that help in assessing the quality of the derived classification model. The second type refers to properties of the model, e. g. complexity of the discovered rules. Although explicitly mentioned in the context of classification, such measures may also be applied in unsupervised learning tasks. Finally, properties of the data, e. g. feature space, distribution, etc., may certainly be applied to both supervised and unsupervised learning.

Performance Measures. In the context of adaptive information filtering, Klinkenberg and Renz evaluate performance measures by conducting experiments using different algorithms and settings [23]. They use accuracy, precision and recall as inputs for an algorithm that dynamically adjusts the size of the time window on the training data. They show that in classification tasks such measures are good indicators for pattern change, whereby recall and precision outperform accuracy when identifying concept changes.

Simpler methods that are based on performance measures are also used by e. g. [20] and [41]. However, it is clear that most of these indicators are not applicable in the context of unsupervised learning. For example, a user may measure the performance of a classifier to see if it should be adapted. On the contrary, in clustering tasks, population drift implies that the contents and shape of a clustering scheme may change, but the notion of a *correct* or *proper* clustering is not well-defined.

Properties of the Data. The DELI Change Detector of Lee et al. uses a sampling technique to detect changes that may affect previously discovered association rules and invokes an incremental miner to modify the patterns as needed [26,27].

Ganti et al. propose the DEMON framework for data evolution and monitoring across the temporal dimension [18]. DEMON focuses on detecting systematic vs. non-systematic changes in the data and on identifying the data blocks (along the time dimension) which have to be processed by the miner in order to extract new patterns. However, the emphasis is on updating the knowledge base by detecting changes in the data, rather than detecting changes in the patterns. The closely related framework FOCUS (of the same group) is designed to compare two datasets and compute an interpretable, qualifiable deviation measure between them [16]. Finally, the CACTUS algorithm exploits summaries upon datasets as the basis of "well-defined" clusters, which can then be discovered by only two passes over each of the datasets under consideration [17].

The works of Ganti et al., when observed as components of a complete framework, are the closest relevant work to our study of pattern evolution and identification of global and local patterns. Although the focus of [16,17,18] is on knowledge alignment, the components could be used as the basis for a temporal modelling of clusterings and the analysis of their evolution. In our pattern evolution method, we take a different approach by starting with a temporal model for arbitrary patterns (in unsupervised learning) and focusing on the underpinnings of their change rather than on their adaptation.

Properties of the Model. The IncrementalDBSCAN algorithm of Ester et al. [13] extends the DBSCAN clustering algorithm by a component that deals with record insertions and their effects on the contents, centroids and borders of clusters. In this approach, there are different types of cluster members; a cluster disappears when all its so-called "strong" members have migrated [13]. In principle, they track the movement of the strong cluster members as new data is added in order to decide when a cluster vanishes.

Aslam et al. formalise clustering as the problem of covering graphs with star-shaped dense subgraphs, enumerate the types of impact a record insertion or deletion may have on the covering graph, and then propose an algorithm that adjusts the covering graph(s) accordingly [3]. Similarly to IncrementalDBSCAN, this algorithm adjusts the clustering scheme whenever a new record is inserted.

2.3   Detection of Interesting Changes

While classification results are usually evaluated with respect to their predictive performance, the evaluation of patterns in unsupervised learning is more challenging. For clustering, the question of a "good" clustering scheme is prominent. For association rules and frequent sequences, a ranking of the patterns on "interestingness" is of paramount importance. When observing pattern changes caused by drifts and shifts of the underlying population, it is equally important to assess the interestingness of the observed changes.

Interestingness criteria, as a sub-category of quality evaluation criteria, are discussed in [32,46]. Tuzhilin et al. elaborate on the criterion of pattern "unexpectedness" towards the beliefs of the human expert [2,9,34,35,40]. Rule unexpectedness is also discussed in [15,45]. The selection of appropriate interestingness measures for association rules is addressed in [28,44,45]. However, most of these studies observe patterns derived over a static population. The temporal aspect of patterns is addressed in [9,11].

Berger and Tuzhilin elaborate on the discovery of interesting repetitions (reappearances) of a pattern across a series of events. Here a pattern is interesting if the ratio of its actual versus its expected occurrences exceeds a given threshold [9]. Pattern discovery is based on temporal predicates, supporting the operators NEXT, BEFORE_k (with k being a given number of events) and UNTIL. The model of Karimi and Hamilton on the discovery of causality relationship among events [21] further delivers a particular form of interesting temporal rules for the context of temporal classification. Interestingness models for sequences of events are further addressed in [42]. However, both works, as well as further studies on simple or complex types of events [33,25] focus on correlations among events belonging to the same rule rather than on correlated rules.

Chakrabarti et al. focus on the temporal properties of potentially associated patterns [11]: they partition the time axis into time slots in such a way, that pairs of association rules co-occurring in an unexpected way are identified.

Closely related to the work of [9,11] are the temporal mining studies of [12,37], where the focus is on the discovery of the maximum valid interval for a rule, subject to statistical constraints. From the viewpoint of temporal mining, the terms "meta-mining" in [1] and "higher-order mining" in [39] have been proposed as labels for the discovery of temporal patterns among conventional patterns occurring along the time axis. Although there is no explicit emphasis on the concept of "interestingness", it is pointed out that a change in the statistical properties of an association rule or a frequent sequence is a phenomenon of potential interest. Our work on the evolution of patterns modelled as temporal objects and subject to changes of content and statistics [5,6,7,8], provides the framework and the heuristics for the discovery of interesting changes.

A model for interesting rule changes across the time axis is proposed in [29]. This model categorises rules into stable rules that exhibit no variation, rules that show a clear trend and semi-stable rules that stand between the other two types. The dataset is partitioned, the partitions are analysed separately and heuristics are used to juxtapose the statistics of the rules across the partitions and assign them to one of the three categories [29]. Our model uses a slightly similar concept of pattern "stability" but combines it with "persistence" and "slope"; it applies different types of heuristics to map patterns across those dimensions.

3 Temporal Management and Monitoring of Patterns

Goal of our "Pattern Monitor" is the observation of pattern evolution across the time-line and the detection of interesting changes in patterns. Unlike pattern alignment techniques, our objective is not the adaptation of patterns as the underlying population shifts or drifts but rather the identification and categorisation of changes according to an interestingness model, so that the mining expert can be notified accordingly.

For our pattern monitoring approach, we consider knowledge discovery as a series of mining sessions over time between which data accumulates. Each session reveals a set of patterns, some of them may be known from previous sessions, while some are new and other patterns may disappear. As opposed to the knowledge alignment techniques discussed in Section 2, known patterns are not updated. Instead, each pattern becomes a self-standing object in the rule base; its statistics become instances describing the pattern at each time point of its existence. Accordingly, we model patterns as temporal objects, shift from interesting patterns to interesting pattern changes and devise heuristics that detect specific types of interesting change.

3.1   A Generic Model for Patterns as Temporal Objects

We represent patterns as temporal objects according to the model in [5]. In this model, mining sessions are initiated at specific timestamps t_i, $i = 0, 1, \ldots, n$, where t_0 is the time point of the first analysis. In the mining session at timestamp t_i, we analyse the dataset D_i collected in the period $t_i - t_{i-1}$, $i \geq 1$. Hence, a pattern ξ is a temporal object with the following signature [3]:

$$\xi = ((ID, query, body, head), \{(period, statistics)\})$$

For association rules, *body* is the rule's antecedent and *head* is the consequent. For frequent sequences, the pattern has the form *body·head*. For clusterings, each individual cluster is a pattern; *body* and *head* are empty. The *query* is the specification of the mining parameters guiding the discovery process. Its syntax depends on the interface of the miner being invoked and is external to our model. It is retained as part of each pattern, though, because the instances of a pattern over time only make sense if they are discovered via the same query.

The system-generated ID connects the invariant part of a pattern with the part that can change at each timestamp. For association rules and frequent sequences, *body* and *head* belong to the invariant part, while the statistics may change from timestamp to timestamp. A change in the pattern's content implies a new pattern. For clusterings, the only invariant parts of a cluster are the cluster identifier returned by the miner and the *query* that generated the clustering: Tracing the same cluster across time is a challenging problem. Presently, we rather perform clustering once and trace the contents of each cluster upon the data accumulated at later timestamps.

For each period, there is one instance of the *statistics* of ξ. These statistics depend on the rule type: For association rules we consider support, confidence and lift, for clusters we record intra-cluster distance and cardinality.

Example 1. An association rule "$A \Rightarrow B$" with support 25% and confidence 90%, discovered during period t_1 by mining query q is represented by $R_{AB} = (ID_{AB}, q, t_1, [support = 0.25, confidence = 0.9], A, B)$.

3.2   Pattern Interestingness

The observation of pattern evolution can be performed in different ways: (i) A new mining session is launched at each timestamp, whereupon the set of discovered patterns is juxtaposed to the sets previously discovered. For already existing patterns, a new instance of their statistics is inserted in the rule-base. A new temporal object is created for each emerged pattern, whereupon the statistic values of former (not observed) instances may be set to null. Alternatively, (ii) only an initial mining session is launched. At each later timestamp, an instance of

---

[3] We use the terms "timestamp" and "period" interchangeably hereafter: t_i corresponds to the period during which dataset D_i was collected, i.e., between timestamps t_{i-1} and $t_i, i \geq 1$.

each existing pattern is created, computing the statistic values from the dataset accumulated over the corresponding time period. An elaboration of the two methods can be found in [4].

As soon as the statistic values of all patterns under observation have been filled up to the current timestamp t_i, pattern monitoring can be performed upon the *lifetime* $t_0 \ldots t_i$ of each pattern. The objective of the pattern monitor is to identify those patterns, which have experienced an "interesting change", and to alert the mining expert accordingly.

We extend the traditional notion of pattern interestingness, which is based on the statistics of the pattern at a time point, by considering the behaviour of a pattern's statistics over its whole lifetime. This allows for the identification of *interesting trends* over otherwise *uninteresting properties*. For example, a pattern showing high values for support and confidence may be obvious from the application point of view. Still, if the support of this pattern exhibits large variations or grows over a long period, this may be of importance to the user. Hence, we define the interestingness of patterns across the following dimensions:

- *Persistence:* Fraction of timestamps in which pattern instances are observed. *Global patterns* are those observable during most of the lifetime of a pattern. Temporarily (in)visible patterns are *local*.
- *Stability:* Variation in the statistics of the pattern's instances.
- *Slope/Trend:* Direction of change in the statistics of the pattern's instances.

Figure 1 shows the relationship among the different dimensions of interestingness for global and local patterns, respectively. As the number of changes increases over time, the pattern becomes more noisy, while a pattern that shows no changes at all is invariant. Depending on the slope of the changes we can separate drifts from shifts. This is true for both global and local patterns, but drifts and/or shifts of the latter refer only to a subset of the population.

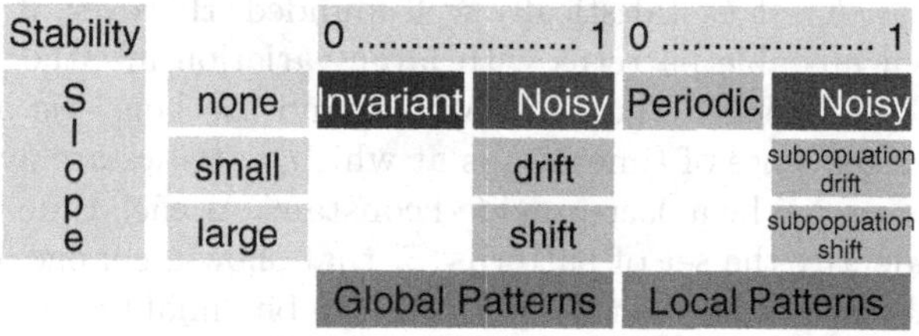

Stability		0 1	0 1
S none	Invariant	Noisy	Periodic	Noisy
l o small	small	drift	subpopuation drift
p e large	large	shift	subpopuation shift
		Global Patterns	Local Patterns

Fig. 1. Interestingness for global and local patterns

We apply the interestingness dimensions among patterns by incorporating them to a pattern monitor that alerts the mining expert whenever (1) formerly global patterns become local, (2) variations of patterns exceed some stability thresholds or (3) the slope of change is beyond a threshold value. In particular, let ξ be a pattern discovered via a mining query q and let $T_s(\xi)$ be the time-series for its statistical property s. Further, let t_i for $i \geq 1$ be the present timestamp, implying that the pattern's lifetime spans the periods from t_0 to t_i. Then:

1. The *persistence* test labels a pattern as "local" if $\frac{|\{t_j | T_s(\xi)[t_j] < \tau_s\}|}{i} > \tau_{global}$. In this test, τ_s is the threshold set upon property s, e.g. the minimum confidence or support of an association rule. This threshold may be set in the mining query q or by the expert upon already discovered patterns. The threshold τ_{global} refers to the fraction of periods, where the pattern should have been observed to be still considered global. If it is set to zero, then a pattern is local as soon as it disappears for one period.

2. The *stability* test labels a pattern ξ as "changed" if $T_s[\xi](t_i)$ is different from past values. We assess changes with the heuristics described below.

3. The *slope* is the angle of T_s at t_i and is computed upon changed patterns. The label "small", resp. "large" depends on the (expert-specified) threshold.

3.3 Detecting Pattern Changes

Presently, we observe stability and slope as closely interrelated properties of patterns. In particular, we focus on distinguishing between unstable patterns and patterns that exhibit trends, i.e. have a non-negligible slope. For this purpose, we use heuristics that monitor the time-series of pattern statistics and raise alerts.

Significance Tester. This heuristic applies a two-tailed binomial test to verify whether an observed change is statistically significant or not: For a pattern ξ and a statistical measure s, we test at each period t_i whether $\xi.s(t_{i-1}) = \xi.s(t_i)$ at a confidence level α. The test is applied upon the data subset D_i accumulated between t_{i-1} and t_i, so that the null hypothesis means that D_{i-1} is drawn from the same population as D_i. Then, for a pattern ξ an alert is raised for each time period t_i at which the null hypothesis is rejected.

The significance test is statistically well-founded. However, it makes no distinction between unstable patterns with large variation in either direction and patterns with clear trends. Therefore, we expand the heuristic as follows: Let $t_0, \ldots, t_n$ be the sequence of timestamps at which statistics are available for the patterns and let $m < n$ be a user-provided constant. At each time period t_i, with $i < n - m$ we identify the set of patterns Ξ_i that show a change at t_i according to the binomial test above. We then perform the binomial test upon the dataset of the aggregated time period from t_i till t_{i+m}. If there is a second significant change within this second period, we test whether the second change cancels the first one. If yes, we label the pattern as "unstable" and report only the first change at t_i. If not, we label the pattern as having a "large slope" and report both changes. In this latter case, we term the change at t_i as a "core alert", which may be used as an indicator of concept drift.

Interval Heuristic. This heuristic partitions the range of values of the statistical property under observation into consecutive intervals and raises alerts when the value observed at a timestamp shifts to another interval than the values in previous timestamps. In particular, for a property s, its value-range $[L_s, U_s]$ is split into k intervals of equal width. [4] Then, for a pattern ξ, this heuristic raises an alert at timestamp t_i if the value of $T_s(\xi)[t_i]$ is at a different interval than the value of $T_s(\xi)[t_{i-1}]$ *and* the absolute difference between the two values is more than an ϵ. The values of k, ϵ are provided by the expert, whereby larger values of k and smaller values of ϵ result in more alerts.

Corridor Heuristic. This heuristic defines a "corridor" around the observed time-series of a pattern. A corridor is an interval within the value-range of the observed statistical property, having the current average of the series as its centre. In particular, let ξ be a pattern and $T_s(\xi)$ the time-series under observation. At timestamp t_i, we compute the mean $m(T_s(\xi)[t_i])$ and the standard deviation $stddev(T_s(\xi)[t_i])$ of the time-series on the basis of the values at the time-points $T_s(\xi)[t_j]$ for $j = i - \tau_{win}, \dots, i$. Here, τ_{win} determines the window length for a sliding window. Then, the width of the corridor at t_i is the interval $I_s(\xi)[t_i]$ spanning one standard deviation $stddev(T_s(\xi)[t_i])$ below and above the mean $m(T_s(\xi)[t_i])$. This heuristic raises an alert if the value $T_s(\xi)$ is outside $I_s(\xi)[t_i]$.

The corridor heuristic exploits the history of a time-series. It is insensitive to oscillations close to the mean and thus can better distinguish between unstable patterns and patterns showing a (temporary) trend. It is obviously sensitive to the values of the standard deviation in the past, so that patterns with large oscillations have also large corridors. This is not necessarily undesirable, since the notion of trend for a pattern does depend on the past variability of the pattern. So, differently from the significance tester, this heuristic may or may not alert for a significant change, depending on how stable the pattern has been thus far.

The corridor heuristic depends on the size of the sliding window. This size τ_{win} may be a *constant* value specified by the expert, *expanding* to cover the whole past, i.e. $\tau_{win} := i$, or *customised*, i.e. computed according to some more sophisticated mechanism that weights past values (e.g. [14]). Since a reliable computation of the mean and standard deviation of a statistical property requires some minimum number of observations, the expert may specify a minimum number of periods τ_{learn}, during which no alerts are raised.

4   Experiments on Maillog Data

We applied our pattern monitoring approach upon association rules over a one year's log of a mailserver. We show some of the findings, namely patterns with interesting changes from the application's perspective.

---

[4] One of L_s, U_s may be a threshold upon s, as posed by the mining query.

4.1 The Mailserver Log and Its Patterns

The dataset comes from the mailserver of a university institute. For each mail, the log contains one entry for the sender (hereafter termed as: *fromline*) and one entry per recipient (hereafter: *toline*). Within one year, approximately 280,000 entries have been accumulated, corresponding to more than 100,000 mails.

The log was cleaned and anonymised by replacing the identifier of sender and recipient(s) by their origin ("institute", "faculty", "external") and adding a unique mail_id to each *fromline* entry and the corresponding *toline* entries. It was then imported into a relational DBMS. The attributes of the *fromline* and of the *toline* were encoded into multi-attribute fields. Association rules discovery was then performed upon the value combinations in the encoded fields.

Table 1 describes the encoding. The first digit determines whether the mail was sent during working hours vs. weekend/holidays. Digits 2 to 4 come from the *fromline* entry and indicate whether the sender belongs to the institute, the institute's faculty or is external. Digits 5 to 7 refer to the location of the recipients. The last two digits denote the number of recipients.

Pos	Description	Pos	Description
1	working hours (1) or not (0)	5	receiver in institute (1) or not (0)
2	sender from institute (1) or not (0)	6	receiver in faculty (1) or not (0)
3	sender from faculty (1) or not (0)	7	receiver external (1) or not (0)
4	sender external (1) or not (0)	8-9	number of recipients (decimal number)

Table 1. Encoding scheme used for the mail log entries

Example 2. For a mail send from an institute member to one institute member, the *fromline* is encoded into 010000001, the single *toline* into 000010001.

For association rules discovery, we used thresholds of 0.025 for support and 0.5 for confidence. All patterns with lift less than 1.0 were eliminated. Over the year, 83 patterns were recorded, but only one of them was global; 73 local patterns were present in less than 75% of the periods. Table 2 gives an overview for the first five periods: For example, there were 13 rules in the first period, 6 of which disappeared in the second period, whereupon 12 new rules were found.

4.2 Discovery of Interesting Changes

We applied the heuristics described in the previous section to study patterns that gave raise to alerts. As expected by the elaboration on the behaviour of the heuristics towards unstable patterns, the alerts of the significance tester were not intuitive, while the corridor heuristic with $\tau_{win} = i$ for each t_i raised interesting

| | | | | number of rules | | | | |
period	items	transactions	itemsets	*total*	*unknown*	*known*	*previous*	*disapp*
1	4058	1485	13	13	13	0	0	0
2	5038	1661	11	19	12	0	7	6
3	6000	1843	11	12	0	4	8	11
4	5454	1808	11	10	1	0	9	3
5	11776	4921	8	8	2	1	5	5

Table 2. Contents of the rule-base in the first five periods.

alerts leading to insights on the behaviour of mail users. Note that the corridor heuristic will raise an alert if the current value of the time series differs stronger from past values than expected.

Fig. 2 shows the time-series of support and confidence for the sole global pattern 000100001 $\Rightarrow$ 000010001. This pattern refers to mails with an external sender and one institute recipient. The frequency of such mails may vary considerably (left part of Figure). Of interest is rather the confidence time-series: It is relatively stable, except for a single dramatic drop in t_{27}. The alert on this change resulted in a juxtaposition of this pattern to the other (local) patterns and to some interesting insights on the subpopulation of the external senders.

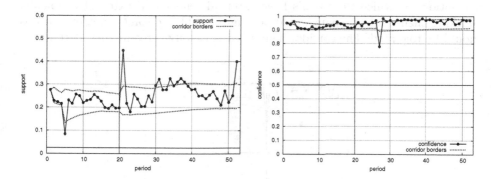

Fig. 2. Time-series of support and confidence for global pattern 000100001 $\Rightarrow$ 000010001

The support time-series of two local patterns are shown in Fig. 3. They refer to mails with a sender from the institute to one, resp. two institute members. The support value crosses the lower boundary several times. The pattern on the right (mails with two recipients) is quite stable. The pattern on the left (mails with one recipient) shows large variability but only one remarkable peak in the period at the end of the winter term, when the preparation of the examinations is done by the institute members. This peak gave raise to an alert by all heuristics. The corridor heuristic gave no further alerts for this rather noisy pattern.

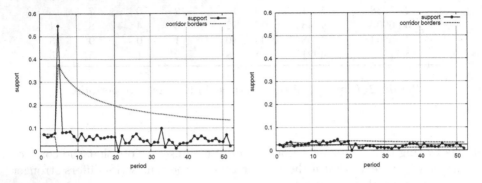

Fig. 3. Support of the patterns $110000001 \Rightarrow 100010001$ and $110000002 \Rightarrow 100010002$

The pattern in Fig. 4 is of particular interest: It refers to mails of an external sender and two recipients, one of them being an institute member. Intuitively, its confidence should be 1, since a mail with an external sender and no internal recipient should not have been recorded at all. The first alert allowed for a better insight in the population: Some institute members use a mail forwarding option, by which their mails have external sender *and* external recipients. Blocking their mails as spam would be unacceptable, so some confidence drops should be tolerated. The corridor heuristic raises only three further alerts.

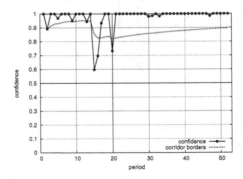

Fig. 4. Confidence time-series of pattern $000100002 \Rightarrow 000010002$

5   Conclusions

In this study, we have elaborated on the subject of pattern evolution as a basis for the detection of interesting changes during a population's lifetime. We have discussed the various research threads that address pattern adaptation to population change and pattern interestingness for stationary and for evolving patterns. In our approach, we have modelled patterns as temporal objects and defined the notion of interestingness upon pattern persistence, stability and slope. We have used simple heuristics to characterise patterns with respect to their persistence and stability and to raise alerts when changes in slope occur.

There are several open issues in the understanding of pattern change. First, a pattern is described by several time-series reflecting its statistics; our current interestingness model considers only one time-series as observation object. Studying the interplay of the time-series of multiple statistical properties will result in better insights into the behaviour of the underlying population. Second, the patterns under observation are interdependent. The understanding of the interdependencies among patterns on overlapping and non-overlapping parts of the population is very important for the interpretation of the population's evolution. Finally, the notion of pattern locality is presently defined as non-persistence across the time axis. A more elaborate model on local phenomena is needed, drawing upon the results of time-series analysis on periodicity and upon findings on correlated changes of different patterns.

References

1. T. Abrahams and John Roddick. Incremental meta-mining from large temporal data sets. In *Proc. of 1st Int. Workshop on Data Warehousing and Data Mining*, pages 41–54, 1998.
2. Gediminas Adomavicius and Alexander Tuzhilin. Discovery of actionable patterns in databases: The action hierarchy approach. In *KDD*, pages 111–114, Newport Beach, CA, Aug. 1997.
3. Javed Aslam, Katya Pelekhov, and Daniela Rus. A practical clustering algorithm for static and dynamic information organization. In *SODA: ACM-SIAM Symposium on Discrete Algorithms (A Conference on Theoretical and Experimental Analysis of Discrete Algorithms)*, pages 51–60, January 1999.
4. Steffan Baron. *Temporale Aspekte entdeckten Wissens: Ein Bezugssystem für die Evolution von Mustern*. PhD thesis, Humboldt University Berlin, 2004. English title: "Temporal Aspects of Discovered Knowledge: A Framework for Pattern Evolution" (On German).
5. Steffan Baron and Myra Spiliopoulou. Monitoring change in mining results. In *Proc. of 3rd Int. Conf. on Data Warehousing and Knowledge Discovery (DaWaK'01)*, Munich, Germany, Sept. 2001.
6. Steffan Baron and Myra Spiliopoulou. Monitoring the results of the KDD process: An overview of pattern evolution. In J. M. Meij, editor, *Dealing with the Data Flood: Mining data, text and multimedia*, chapter 6, pages 845–863. STT Netherlands Study Center for Technology Trends, The Hague, Netherlands, Apr. 2002.

7. Steffan Baron and Myra Spiliopoulou. Monitoring the evolution of web usage patterns. In *Proc. of 1st European Web Mining Forum (EWMF'03)*, Cavtat, Slovenia, Sept. 2003. to appear in the Workshop Proceedings as Springer LNCS/LNAI series (due Summer 2004).
8. Steffan Baron, Myra Spiliopoulou, and Oliver Günther. Efficient monitoring of patterns in data mining environments. In *Proc. of 7th East-European Conf. on Advances in Databases and Inf. Sys. (ADBIS'03)*, LNCS, pages 253–265. Springer, Sept. 2003.
9. Gideon Berger and Alexander Tuzhilin. Discovering unexpected patterns in temporal data using temporal logic. In Opher Etzion, Sushil Jagodia, and Suryanarayana Sripada, editors, *Temporal Databases: Research and Practice*, LNCS 1399, pages 281–309. Springer-Verlag Berlin, Heidelberg, 1998.
10. John Case, Sanjay Jain, Susanne Kaufmann, Arun Sharma, and Frank Stephan. Predictive Learning Models for Concept Drift. *Theoretical Computer Science*, 268(2):323–349, October 2001.
11. Soumen Chakrabarti, Sunita Sarawagi, and Byron Dom. Mining Surprising Patterns Using Temporal Description Length. In Ashish Gupta, Oded Shmueli, and Jennifer Widom, editors, *VLDB'98*, pages 606–617, New York City, NY, August 1998. Morgan Kaufmann.
12. Xiaodong Chen and Ilias Petrounias. Mining temporal features in association rules. In Jan Zytkow and Jan Rauch, editors, *Proc. of 3rd European Conf. on Principles and Practice of Knowledge Discovery in Databases PKDD'99*, number 1704 in LNAI, pages 295–300, Prague, Czech Republic, Sept. 1999. Springer Verlag.
13. Martin Ester, Hans-Peter Kriegel, Jörg Sander, Michael Wimmer, and Xiaowei Xu. Incremental Clustering for Mining in a Data Warehousing Environment. In *Proceedings of the 24th International Conference on Very Large Data Bases*, pages 323–333, New York City, New York, USA, August 1998. Morgan Kaufmann.
14. Wei Fan. Systematic Data Selection to Mine Concept-Drifting Data Streams. In *Proc. of 10th ACM SIGKDD Int. Conf. on Knowledge Discovery and Data Mining (KDD 2004)*, pages 128–137, Seattle, Washington, USA, August 2004. ACM Press.
15. Alex Alves Freitas. On Objective Measures of Rule Surprisingness. In Jan M. Zytkow and Mohamed Quafafou, editors, *Principles of Data Mining and Knowledge Discovery, Proceedings of the Second European Symposium, PKDD'98*, Nantes, France, 1998. Springer.
16. Venkatesh Ganti, Johannes Gehrke, and Raghu Ramakrishnan. A Framework for Measuring Changes in Data Characteristics. In *Proceedings of the 18th ACM SIGACT-SIGMOD-SIGART Symposium on Principles of Database Systems*, pages 126–137, Philadelphia, Pennsylvania, May 1999. ACM Press.
17. Venkatesh Ganti, Johannes Gehrke, and Raghu Ramakrishnan. CACTUS: Clustering categorical data using summaries. In *Proc. of 5th ACM SIGKDD Int. Conf. on Knowledge Discovery and Data Mining (KDD '99)*, pages 73–83, San Diego, CA, Aug. 1999. ACM Press.
18. Venkatesh Ganti, Johannes Gehrke, and Raghu Ramakrishnan. DEMON: Mining and Monitoring Evolving Data. In *Proceedings of the 15th International Conference on Data Engineering*, pages 439–448, San Diego, California, USA, February 2000. IEEE Computer Society.
19. Gunter Grieser. Hypothesis assessments as guidance for incremental and meta-learning. In J. Keller and C. Girard-Carrier, editors, *Proc. 11th European Conference on Machine Learning, Workshop on Meta Learning: Building Automatic Advice Strategies for Model Selection and Method Combination*, pages 97–108, May 2000.

20. Geoff Hulten, Laurie Spencer, and Pedro Domingos. Mining Time-Changing Data Streams. In Foster Provost and Ramakrishnan Srikant, editors, *Proceedings of the 7th ACM SIGKDD International Conference on Knowledge Discovery and Data Mining*, pages 97–106, New York, August 2001. ACM Press.

21. Kamran Karimi and Howard J. Hamilton. Distinguishing causal and acausal temporal relations. In *Proc. of 7th Pacific-Asia Conf. PAKDD2003*, pages 234–240, Seoul, Korea, April/May 2003. Springer.

22. Ralf Klinkenberg and Thorsten Joachims. Detecting Concept Drift with Support Vector Machines. In Pat Langley, editor, *Proceedings of the 17th International Conference on Machine Learning*, pages 487–494, Stanford, USA, 2000. Morgan Kaufmann Publishers, San Francisco, USA.

23. Ralf Klinkenberg and Ingrid Renz. Adaptive Information Filtering: Learning in the Presence of Concept Drift. In *Workshop on Learning for Text Categorization at 15th National Conference on Artificial Intelligence (AAAI-98)*, 1998.

24. Ivan Koychev. Tracking changing user interests through prior-learning of context. *Lecture Notes in Computer Science*, 2347:223–??, 2002.

25. Anni Lau, Siew Siew Ong, Ashesh Mahidadia, Achim Hoffmann, Johanna Westbrook, and Tatjana Zrimec. Mining patterns of dyspepsia symptoms across time points using constraint association rules. In *Proc. of 7th Pacific-Asia Conf. PAKDD2003*, pages 124–135, Seoul, Korea, April/May 2003. Springer.

26. Sau Dan Lee and David Wai-Lok Cheung. Maintenance of Discovered Association Rules: When to update? In *ACM-SIGMOD Workshop on Data Mining and Knowledge Discovery (DMKD-97)*, Tucson, Arizona, May 1997.

27. S.D. Lee, D.W. Cheung, and B. Kao. Is Sampling Useful in Data Mining? A Case in the Maintenance of Discovered Association Rules. *Data Mining and Knowledge Discovery*, 2(3):233–262, September 1998.

28. Bing Liu, Wynne Hsu, and Shu Chen. Using General Impressions to Analyze Discovered Classification Rules. In *Proceedings of the Third International Conference on Knowledge Discovery and Data Mining*, pages 31–36, Newport Beach, USA, August 1997. AAAI Press.

29. Bing Liu, Yiming Ma, and Ronnie Lee. Analyzing the interestingness of association rules from the temporal dimension. In *IEEE International Conference on Data Mining (ICDM-2001)*, pages 377–384, Silicon Valley, USA, November 2001.

30. Katharina Morik and Hanna Köpcke. Features for Learning Local Patterns in Time-Stamped Data. In Katharina Morik, Jean-Francois Boulicaut, and Arno Siebes, editors, *Detecting Local Patterns*, LNCS. Springer, Berlin, Heidelberg, New York, 2005. to appear.

31. Katharina Morik and Stefan Rüping. A Multistrategy Approach to the Classification of Phases in Business Cycles. *Lecture Notes in Computer Science*, 2430:307–319, 2002.

32. Gholamreza Nakhaeizadeh and Alexander Schnabl. Development of multi-criteria metrics for the evaluation of data mining algorithms. In *KDD'97*, pages 37–42, Newport Beach, CA, Aug. 1997. AAAI Press.

33. Anny Ng and Ada Wai chee Fu. Mining frequent episodes for relating financial events and stock trends. In *Proc. of 7th Pacific-Asia Conf. PAKDD2003*, pages 27–39, Seoul, Korea, April/May 2003. Springer.

34. B. Padmanabhan, S. Sen, A. Tuzhilin, N. White, and R. Stein. The identification and satisfaction of consumer analysis-driven information needs of marketers on the www. *European Journal of Marketing, Special Issue on Marketing in Cyberspace*, 32, 1998.

35. Balaji Padmanabhan and Alexander Tuzhilin. A belief-driven method for discovering unexpected patterns. In *KDD'98*, pages 94–100, New York City, NY, Aug. 1998.
36. Michael Pazzani and Daniel Billsus. Learning and revising user profiles: The identification of interesting web sites. *Machine Learning (Kluwer Academic Publishers)*, 27:313–331, 1997.
37. Michal Pěchouček, Olga Štěpánková, and Petr Mikšovský. Maintenance of Discovered Knowledge. In *Proceedings of the 3rd European Conference on Principles of Data Mining and Knowledge Discovery*, Lecture Notes in Computer Science, pages 476–483, Prague, Czech Republic, September 1999. Springer.
38. Ruggero G. Pensa and Jean-Francois Boulicaut. Boolean property encoding for local set pattern discovery: an application to gene expression data analysis. In Katharina Morik, Jean-Francois Boulicaut, and Arno Siebes, editors, *Detecting Local Patterns*, LNCS. Springer, Berlin, Heidelberg, New York, 2005. to appear.
39. John F. Roddick and Myra Spiliopoulou. A survey of temporal knowledge discovery paradigms and methods. *IEEE Trans. of Knowledge and Data Engineering*, Aug. 2002.
40. Avi Silberschatz and Alexander Tuzhilin. What makes patterns interesting in knowledge discovery systems. *IEEE Trans. on Knowledge and Data Eng.*, 8(6):970–974, Dec. 1996.
41. W. Nick Street and YongSeog Kim. A Streaming Ensemble Algorithm (SEA) for Large-Scale Classification. In Foster Provost and Ramakrishnan Srikant, editors, *Proceedings of the 7th ACM SIGKDD International Conference on Knowledge Discovery and Data Mining*, pages 377–388, New York, August 2001. ACM Press.
42. Xingzhi Sun, Maria E. Orlowska, and Xiaofang Zhou. Finding event-oriented patterns in long temporal seuquences. In *Proc. of 7th Pacific-Asia Conf. PAKDD2003*, pages 15–26, Seoul, Korea, April/May 2003. Springer.
43. Nadeem Ahmed Syed, Huan Liu, and Kah Kay Sung. Handling Concept Drifts in Incremental Learning with Support Vector Machines. In Surajit Chaudhuri and David Madigan, editors, *Proceedings of the 5th ACM SIGKDD International Conference on Knowledge Discovery and Data Mining*, pages 317–321, San Diego, USA, August 1999. ACM Press.
44. Pang-Ning Tan and Vipin Kumar. Interestingness Measures for Association Patterns: A Perspective. In *Workshop on Post Processing in Machine Learning and Data Mining at 6th ACM SIGKDD Int'l Conf. on Knowledge Discovery and Data Mining*, Boston, USA, August 2000.
45. Pang-Ning Tan, Vipin Kumar, and Jaideep Srivastava. Selecting the Right Interestingness Measure for Association Patterns. In *Proc. of the 8th ACM SIGKDD Int'l Conf. on Knowledge Discovery and Data Mining*, 2002.
46. Michalis Vazirgiannis, Maria Halkidi, and Dimitrios Gunopoulos. *Uncertainty Handling and Quality Assessment in Data Mining*. Springer, 2003.
47. Dwi H. Widyantoro, Thomas R. Ioerger, and John Yen. An adaptive algorithm for learning changes in user interests. In *Proc. of CIKM'99*, pages 405–412, Kansas City, MO, Nov. 1999. ACM.

Undirected Exception Rule Discovery as Local Pattern Detection

Einoshin Suzuki

Electrical and Computer Engineering, Yokohama National University, Japan
suzuki@ynu.ac.jp

Abstract. In this paper, we give an interpretation of our undirected exception rule discovery as local pattern detection and introduce some of our endeavors. Our undirected exception rule discovery outputs a set of rule pairs, each of which represents a pair of strong rule and its exception rule. A local pattern is defined as a pattern which deviates from a global model, and can be considered to correspond to our exception rule if the global model corresponds to our strong rule. Several attempts for undirected exception rule discovery are introduced in the context of local pattern detection. Our results mainly concern interestingness measure, algorithmic issues, noise modeling, and performance evaluation.

1 Introduction

Traditionally, researchers in KDD (Knowledge Discovery in Databases) were seeking to find global models that explain most of the examples in the data set. This applies to rule discovery, where a rule [1, 10] is a statement of a regularity in the form of "if *premise* then *conclusion*". Most of the researchers in rule discovery have focused on discovering a set of strong rules, each of which is a description of a regularity for numerous objects with few counterexamples, from a data set. However, it has been pointed out that such a strong rule is often uninteresting, because it typically represents a well-known fact in the domain [6, 7, 9, 11, 12, 13, 14, 15, 16, 17, 18, 19, 20, 21, 22, 24].

In 2002, David Hand proposed to decompose an explanation of data into a background model, local patterns, and random noise [2]. In 2004, The definition of local patterns was intensively discussed at the Dagstuhl Seminar 04161, which brought the followings as an agreement [2].

- Local patterns cover small parts of the data space.
- local patterns deviate from the distribution of the population of which they are part.
- Local patterns show some internal structure.

Local patterns can be interesting as targets of discovery since they deviate from their background model thus they might represent unknown facts in the domain.

The above decomposition fits the definition of exception rules [4, 6, 7, 8, 9, 11, 12, 13, 14, 15, 16, 17, 18, 19, 20, 21, 22, 24], each of which is a deviation

K. Morik et al. (Eds.): Local Pattern Detection, LNAI 3539, pp. 207–216, 2005.

from a strong rule. A strong rule is discovered from the entire data, its exception rule mostly explains the counterexamples of the strong rule, and the remaining examples which are neither explained by the strong rule nor the exception rule are ignored as random noise. For instance, a strong rule "using a seat belt is safe" explains most of the examples, its exception rule "using a seat belt is risky for a child" explains a part of counterexamples of the strong rule, and children who are safe by using seat belts are extremely small in number thus are ignored. It should be noted that an exception rule is known to be possibly beneficial since it can differ from a basis for people's daily activity.

Discovery methods for exception rules can be divided into two approaches from the viewpoint of background knowledge[1]. In a directed approach [6, 8, 9], a method is first provided with background knowledge typically in the form of rules, then the method obtains exception rules each of which deviates from these rules[2]. In an undirected approach [4, 7, 11, 12, 13, 14, 15, 16, 17, 18, 19, 20, 21, 22, 24], on the other hand, no background knowledge is provided. The target of discovery is typically a set of rule pairs each of which consists of an exception rule and its corresponding strong rule. In the framework of local pattern detection, the direct approach corresponds to providing information of the model while the undirected approach seeks for the models and the local patterns simultaneously. If the objective of local pattern detection concerns discovery of unknown facts in the domain, the undirected approach is more appropriate than the directed approach.

Despite of its importance, one of the major difficulties for an undirected approach corresponds to its time complexity. Rule discovery is time-consuming unless it resorts to fast heuristic search. Compared with the directed approach, the undirected approach suffers from extra search for strong rules. Let the number of examples in a data set and the number of conditions in a premise be n and M respectively, then the time complexity of single-rule discovery is typically $\Omega(n^{M+1})$, while rule-pair discovery requires $\Omega(n^{2M+1})$. Note that the former searches for combinations of M conditions in the premise and the conclusion while the latter has extra M conditions in the premise of the exception rule. Therefore, algorithmic issues for making the discovery algorithm time-efficient are important. Other important issues concern interestingness measure, noise modeling, and performance evaluation. In this paper, we summarize a part of our results for these issues.

2   Description of the Problem

We assume that an example e_i is a description about an object stored in a data set in the form of a record, and a data set contains n examples $e_1, e_2, \cdots, e_n$. An example e_i is represented by a tuple $< y_{i1}, y_{i2}, \cdots, y_{im} >$ where $y_{i1}, y_{i2}, \cdots, y_{im}$

---

[1] According to Silberschatz and Tuzhilin, these approaches can be named as subjective and objective [9] respectively.

[2] Several methods such as [8] provide additional search to find more interesting rules from the exception rules.

are values for m discrete attributes. Here, a continuous attribute is supposed to be converted to a nominal attribute using an existing method such as presented in [3]. An event representing a value assignment to an attribute will be called an atom.

We define a conjunction rule as a rule of which premise is represented by a conjunction of atoms, and of which conclusion is a single atom.

$$Y_\mu \to x \tag{1}$$

where $Y_\mu \equiv y_1 \wedge y_2 \wedge \cdots \wedge y_\mu$ is a conjunction of atoms and x is a single atom. We assume that atoms y_1, y_2, $\cdots$, y_μ, x have different attributes. In this paper, we mainly consider the problem of finding a set of rule pairs each of which consists of an exception rule associated with a strong rule. Suppose a strong rule is represented by "if Y_μ then x", where $Y_\mu \equiv y_1 \wedge y_2 \wedge \cdots \wedge y_\mu$ is a conjunction of atoms and x is a single atom. Let $Z_\nu \equiv z_1 \wedge z_2 \wedge \cdots \wedge z_\nu$ be a conjunction of atoms and x' be a single atom which has the same attribute but a value different to the atom x, then the exception rule is represented by "if Y_μ and Z_ν then x'". The discovered pattern in our approach is, therefore, represented by a rule pair $r(x, x', Y_\mu, Z_\nu)$, where M is a user-specified parameter for the maximum number of atoms in a premise.

$$r(x, x', Y_\mu, Z_\nu) \equiv \begin{cases} Y_\mu & \to x \\ Y_\mu \wedge Z_\nu & \to x' \end{cases} \tag{2}$$
$$\mu, \ \nu \leq M$$

Our objective is to discover a set of (possibly) interesting rule pairs from a data set. The set is specified by assuming either an evaluation criterion or a set of constraints.

From the viewpoint of local pattern detection, our undirected exception rule discovery can be considered as a general case of subgroup discovery [5], which represents the case of $Y_\mu = \phi$ in (2). In this case, a rule pair degenerates to $(x, Z_\nu \to x')$ thus the problem can be regarded as search for Z_ν which defines a subgroup for a different conclusion x'. We believe that our undirected exception rule discovery can be considered as a general case of subgroup discovery since the former has extra search for (Y_μ, x, x'). Though we don't deal with the case of subgroup discovery, it would be straightforward to include it in our methods.

3   Methods for Undirected Exception Rule Discovery

3.1   MEPRO with Its Interestingness Measure and Branch-and-Bound Method

Our rule-pair discovery method MEPRO [11] represents the first method for undirected discovery of exception rules. From the viewpoint of local pattern detection, its interestingness measure, which balances the measure for the model and the measure for the local pattern, deserves attention. Moreover, its branch-and-bound method represents an effective example of algorithmic issues for making the discovery algorithm time-efficient.

MEPRO is based on a rule discovery system ITRULE [10]. The essential of ITRULE lies in its interestingness measure J, which corresponds to the quantity $J(x;\ y)$ of information compressed by a rule $y \to x$.

$$J(x;\ y) = \Pr(y)\ j(x;\ y) \tag{3}$$

$$\text{where } j(x;\ y) = \Pr(x|y)\log_2 \frac{\Pr(x|y)}{\Pr(x)} + \Pr(\overline{x}|y)\log_2 \frac{\Pr(\overline{x}|y)}{\Pr(\overline{x})} \tag{4}$$

We defined our measure of interestingness of a rule pair as a product ACEP (x, Y_μ, x', Z_ν) of J-measure of a strong rule and J-measure of an exception rule [11]. Our motivation was to obtain rule pairs each of which consists of rules with large J-measure values. We have proved that $J(x; Y_\mu) + J(x'; Y_\mu \wedge Z_\nu)$ is inappropriate as an evaluation index since it is dominated by one of $J(x; Y_\mu)$ and $J(x'; Y_\mu \wedge Z_\nu)$ when it is large [11].

$$\text{ACEP}(x, Y_\mu, x', Z_\nu) \equiv J(x; Y_\mu)J(x'; Y_\mu \wedge Z_\nu) \tag{5}$$

We have then proposed a discovery algorithm which generates K rule pairs, where K is a user-specified parameter. In the algorithm, a discovery task is viewed as a search problem, in which a node of a search tree represents a rule pair $r(x, x', Y_\mu, Z_\nu)$. A depth-first search method with maximum depth D is employed to traverse this tree. We begin by adding atoms in the conclusions, and then add atoms in the premises. K Rule pairs which have the highest scores for $\text{ACEP}(x, Y_\mu, x', Z_\nu)$ are updated during the search. Let $\mu = 0$ and $\nu = 0$ represent the state in which $Y_\mu = Z_\nu = \phi$, then we define that $\mu = \nu = 0$ holds in a node of depth 1, and as the depth increases by 1, an atom is added to the premise of the general or exceptional rule. A node of depth 2 is assumed to satisfy $\mu = 1$ and $\nu = 0$; a node of depth 3, $\mu = \nu = 1$; and a node of depth $l\ (\geq 4)$, $\mu + \nu = l - 1\ (\mu, \nu \geq 1)$. Therefore, a descendant node represents a rule pair $r(x, x', Y_{\mu'}, Z_{\nu'})$ where $\mu' \geq \mu$ and $\nu' \geq \nu$.

Note that, during the search process, we can safely prune a subtree of the search tree if all the rule pairs in the subtree have lower scores for $\text{ACEP}(x, Y_\mu, x', Z_\nu)$ than the K-th highest score exhibited by the discovered rule pairs. This idea is the motivation behind the branch-and-bound method, which requires an upper-bound of $\text{ACEP}(x, Y_\mu, x', Z_\nu)$ of a subtree of the search tree. According to the following theorem, an upper-bound exists for the ACEP of the rule pairs in the search tree [11].

Theorem 1. Let $\text{H}(\alpha) \equiv [\alpha/\{(1 + \alpha)\Pr(\overline{x})\}]^{2\alpha}/\{(1 + \alpha)\Pr(x)\}$, α_1 and α_2 satisfy $\text{H}(\alpha_1) > 1 > \text{H}(\alpha_2)$, and $ACEP = \text{ACEP}(x, Y_{\mu'}, x', Z_{\nu'})$. If $\text{H}(\Pr(x', Y_\mu, Z_\nu)/\Pr(x, Y_\mu)) < 1$ then,

$$ACEP < \alpha_2 \Pr(x, Y_\mu)^2 \left\{ \log_2 \left(\frac{1}{1 + \alpha_1} \frac{1}{\Pr(x)} \right) + \alpha_1 \log_2 \left(\frac{\alpha_1}{1 + \alpha_1} \frac{1}{\Pr(\overline{x})} \right) \right\}$$

$$\cdot \log_2 \frac{1}{\Pr(x')}$$

else

$$ACEP \leq \left\{ \Pr(x, Y_\mu) \log_2 \left(\frac{\mathrm{p}(x, Y_\mu)}{\Pr(x, Y_\mu) + \Pr(x', Y_\mu, Z_\nu)} \frac{1}{\Pr(x)} \right) + \Pr(x', Y_\mu, Z_\nu) \right.$$

$$\left. \cdot \log_2 \left(\frac{\Pr(x', Y_\mu, Z_\nu)}{\mathrm{p}(x, Y_\mu) + \Pr(x', Y_\mu, Z_\nu)} \frac{1}{\Pr(\overline{x})} \right) \right\} \Pr(x', Y_\mu, Z_\nu) \log_2 \frac{1}{\Pr(x')}$$

This upper bound was employed in our approach for a branch-and-bound method which guarantees the optimal solution and is expected to be time-efficient. In other words, our branch-and-bound method typically speeds up search without changing the discovery outcome.

We have also introduced probabilistic constraints for eliminating rule pairs each of which has a large $\Pr(x'|Z_\nu)$ [12]. We have also considered unexpectedness from a different perspective and proposed a novel probabilistic criterion which mainly considers the number of counter-examples [14].

3.2 PADRE with Its Noise Distinction

In the next method PADRE [13, 20], we mainly pursued the problem of distinguishing local patterns from noise. Intuitively, an exception rule which explains "few" examples can be considered as noise rather than a local pattern. We have proposed a method based on simultaneous estimation of true probabilities as an analytical solution to this problem. In PADRE, on the other hand, the interestingness measure is simplified to threshold specification and as the result, the branch-and-bound method is substituted by a pruning method.

In rule discovery, generality and accuracy can be considered as frequently-used criteria for evaluating the goodness of a rule. In case of a conjunction rule $Y_\mu \rightarrow x$, these two criteria correspond to the probability $\Pr(Y_\mu)$ of the premise and the conditional probability $\Pr(x|Y_\mu)$ of the conclusion given the premise respectively [10]. Similar to [1], we specify two minimum thresholds θ_1^S and θ_1^F for generality and accuracy of the strong rule respectively. Two thresholds θ_2^S and θ_2^F are also specified for generality and accuracy of the exception rule respectively.

Consider the case in which the accuracy of a rule $Z_\nu \rightarrow x'$, which we call a reference rule, is large. In such a case, an exception rule can be considered as expected since it can be easily guessed from this rule. In order to obtain truly unexpected exception rules, we specify a maximum threshold θ_2^I for the accuracy of a reference rule.

We then proposed the method PADRE in which we specify thresholds θ_1^S, θ_1^F, θ_2^S, θ_2^F, θ_2^I for probabilistic criteria of a rule pair. Since a rule pair discovered from 10,000 examples exhibits different reliability from another rule pair discovered from 100 examples, it is inappropriate to use a ratio $\widehat{\Pr}(\cdot)$ in a data set as a probabilistic criterion. Therefore, we considered a true probability $\Pr(\cdot)$ for each probabilistic criterion, and obtained a set of rule pairs each of which satisfies discovery conditions with the significance level δ [13, 20]. In the following, $\mathrm{MIN}(a, b)$ and $\mathrm{MAX}(a, b)$ represent the smaller one and the larger one of a and b respectively.

$$\Pr[\ \Pr(Y_\mu) \geq \theta_1^S,\ \Pr(x|Y_\mu) \geq \text{MAX}(\theta_1^F, \widehat{\Pr}(x)),\ \Pr(Y_\mu Z_\nu) \geq \theta_2^S,$$
$$\Pr(x'|Y_\mu Z_\nu) \geq \text{MAX}(\theta_2^F, \widehat{\Pr}(x')), \Pr(x'|Z_\nu) \leq \text{MIN}(\theta_2^I, \widehat{\Pr}(x'))\ ] \geq 1 - \delta$$
$$\text{(6)}$$

Calculating (6) is difficult due to two reasons. First, obtaining a value of a true probability requires assumptions. Second, calculating (6) for a rule pair numerically is time-consuming since (6) contains five true probabilities. PADRE overcomes these difficulties by obtaining analytical solutions based on simultaneous estimation of true probabilities. Let the number of examples in the data set be n, and $(n\Pr(xY_\mu Z_\nu),\ n\Pr(x'Y_\mu Z_\nu),\ n\Pr(\overline{x}\overline{x'}Y_\mu Z_\nu),\ n\Pr(xY_\mu \overline{Z_\nu}),\ n\Pr(\overline{x}Y_\mu \overline{Z_\nu}),\ n\Pr(\overline{x'Y_\mu}Z_\nu),\ n\Pr(x'\overline{Y_\mu}Z_\nu))$ follow a multi-dimensional normal distribution, then (6) is equivalent to (7) - (11) [13, 20].

$$G(Y_\mu,\ \delta,\ k)\widehat{\Pr}(Y_\mu) \geq \theta_1^S \tag{7}$$

$$F(Y_\mu,\ x,\ \delta,\ k)\widehat{\Pr}(x|Y_\mu) \geq \theta_1^F \tag{8}$$

$$G(Y_\mu Z_\nu,\ \delta,\ k)\widehat{\Pr}(Y_\mu Z_\nu) \geq \theta_2^S \tag{9}$$

$$F(Y_\mu Z_\nu,\ x',\ \delta,\ k)\widehat{\Pr}(x'|Y_\mu Z_\nu) \geq \theta_2^F \tag{10}$$

$$F'(Z_\nu,\ x',\ \delta,\ k)\widehat{\Pr}(x'|Z_\nu) \leq \theta_2^I \tag{11}$$

where $$G(a,\ \delta,\ k) \equiv 1 - \beta(\delta, k)\sqrt{\frac{1 - \widehat{\Pr}(a)}{n\widehat{\Pr}(a)}} \tag{12}$$

$$F(a,\ b,\ \delta,\ k) \equiv 1 - \beta(\delta, k)\varphi(a, b) \tag{13}$$
$$F'(a,\ b,\ \delta,\ k) \equiv 1 + \beta(\delta, k)\varphi(a, b)$$

$$\varphi(a,\ b) \equiv \sqrt{\frac{\widehat{\Pr}(a) - \widehat{\Pr}(a, b)}{\widehat{\Pr}(a, b)\{(n + \beta(\delta, k)^2)\widehat{\Pr}(a) - \beta(\delta, k)^2\}}} \tag{14}$$

Here $\beta(\delta, k)$ represents a positive value which is related to the confidence region and is obtained by numerical integration [13, 20]. In PADRE, we assume that a true probability among $\Pr(xY_\mu Z_\nu),\ \Pr(x'Y_\mu Z_\nu),\ \Pr(\overline{x}\overline{x'}Y_\mu Z_\nu),\ \Pr(xY_\mu \overline{Z_\nu}),$ $\Pr(\overline{x}Y_\mu \overline{Z_\nu}),\ \Pr(\overline{x'Y_\mu}Z_\nu),\ \Pr(x'\overline{Y_\mu}Z_\nu)$ is equal to 0 if the corresponding estimated probability is equal to 0. Therefore, the number k' of true probabilities each of which is satisfied by at least an example in the data set determines the dimensionality of the ellipsoid which corresponds to the confidence region of $(n\Pr(xY_\mu Z_\nu),\ n\Pr(x'Y_\mu Z_\nu),\ n\Pr(\overline{x}\overline{x'}Y_\mu Z_\nu),\ n\Pr(xY_\mu \overline{Z_\nu}),\ n\Pr(\overline{x}Y_\mu \overline{Z_\nu}),$ $n\Pr(\overline{x'Y_\mu}Z_\nu),\ n\Pr(x'\overline{Y_\mu}Z_\nu))$. In (7) - (14), $k = k' - 1$. We have also proposed an efficient discovery algorithm based on pruning.

4   Evaluation with Real Data Sets

In the Dagstuhl Seminar 04161, the participants recognized that benchmark data for evaluating methods for local pattern detection were missing. Our evaluation, which we present in this section, for undirected exception rule discovery with real data sets might suggest a solution to this problem.

We participated in a data mining contest with the meningitis data set [17]. The data set consists of 140 patients each of whom is described by 38 attributes and has been made public as a benchmark problem to the data mining community. Since the data set is relatively small, we used a modified version of PADRE without noise distinction i.e. we used the following as evaluation criterion.

$$\Pr(Y_\mu) \geq \theta_1^S, \ \Pr(x|Y_\mu) \geq \mathrm{MAX}(\theta_1^F, \widehat{\Pr}(x)), \ \Pr(Y_\mu Z_\nu) \geq \theta_2^S,$$

$$\Pr(x'|Y_\mu Z_\nu) \geq \mathrm{MAX}(\theta_2^F, \widehat{\Pr}(x')), \Pr(x'|Z_\nu) \leq \mathrm{MIN}(\theta_2^I, \widehat{\Pr}(x')) \qquad (15)$$

Since specification of thresholds θ_1^S, θ_1^F, θ_2^S, θ_2^F, θ_2^I can be a laborious task, we used a method which dynamically adjusts the values of the thresholds [15].

Our method has discovered 169 rule pairs from a pre-processed version of this data set [17]. These rule pairs were inspected by Dr. Tsumoto, who is a domain expert, and each rule pair was assigned a five-rank score for the following evaluation criteria each of which was judged independently.

- validness: the degree that the discovered pattern fits domain knowledge
- novelty: the degree that the discovered pattern does not exist in domain knowledge
- usefulness: the degree that the discovered pattern is useful in the domain
- unexpectedness: the degree that the discovered pattern partially contradicts domain knowledge

For the scores, five and one represent the best score and the worst score respectively. We show the results classified by the attributes in the conclusions in Table 1.

From the Table, we see that the average scores of the discovered rule pairs are high for several attributes in the conclusions. We inspected these rule pairs by grouping them with respect to the attribute in the conclusion, and found that these attributes can be classified into four categories. The first category represents attributes with the lowest scores, and includes CULTURE, C_COURSE, and RISK. We consider that attributes in this category cannot be explained with this data set, and investigation on them requires further information on other attributes. The second category represents attributes with higher scores for validness and usefulness, and includes FOCAL, LOC_DAT, and Diag2. We consider that attributes in this category can be explained with this data set, and has been well investigated probably due to their importance in this domain. We regard them as one of important targets in discovery although one will often rediscover conventional knowledge. The third category represents attributes with approximately equivalent scores, and includes CT_FIND, EEG_FOCUS, and Course (G). We consider that attributes in this category can be explained with this data set, and has not been investigated well in spite of their importance in this domain. We regard them as one of the most important targets in discovery. The fourth category represents attributes with higher scores for novelty and unexpectedness, and includes CULT_FIND, KERNIG, and SEX. We

Table 1. Average performance of the proposed method with respect to attributes in the conclusion. The column "#" represents the number of discovered rule pairs.

attribute	#	validness	novelty	unexpectedness	usefulness
(all)	169	2.9	2.0	2.0	2.7
CULTURE	2	1.0	1.0	1.0	1.0
C_COURSE	1	1.0	1.0	1.0	1.0
RISK	1	1.0	1.0	1.0	1.0
FOCAL	18	3.1	2.2	2.7	3.0
LOC_DAT	11	2.5	1.8	1.8	2.5
Diag2	72	3.0	1.1	1.1	2.6
CT_FIND	36	3.3	3.0	3.0	3.2
EEG_FOCUS	11	3.0	2.9	2.9	3.3
Course (G)	8	1.8	2.0	2.0	1.8
CULT_FIND	4	3.3	4.0	4.0	3.5
KERNIG	4	2.0	3.0	3.0	2.0
SEX	1	2.0	3.0	3.0	2.0

consider that attributes in this category can be explained with this data set, but has been somewhat ignored. We consider that investigating these attributes using discovered rule sets can lead to interesting discoveries which might reveal unknown mechanisms in this domain in spite of their apparent low importance.

As Dr. Tsumoto admits, our success is due to the fact that the structure of a rule pair is useful for discovery of interesting patterns. According to him, our method discovered the most interesting results in the data mining contest [23].

Our method has been also applied to 1994 bacterial test data set (20,919 examples, 135 attributes, 2 classes) [18]. We have found that we need to consider distribution of attribute values and cause and effect relationships in order to discover interesting patterns from the data set. However, this application shows that our method is adequate in terms of efficiency in exception rule mining from a relatively large-scale data set.

5   Concluding Remarks

In this paper, we have introduced our undirected exception rule discovery from the viewpoint of local pattern detection. MEPRO shows an effective solution to the interestingness measure problem and the time inefficiency problem. PADRE, on the other hand, shows a solution to how to distinguish local patterns from noise. Our endeavor in a data mining contest might suggest a solution to the problem of performance evaluation.

In local pattern detection, a global model and a local pattern typically adopt different representations. Our undirected exception rule discovery represents each of them as a rule thus might deviate from our intuition on local pattern detection. A global model typically explains most of the data while our strong rule, which corresponds to the global model in our framework, explains a large number of examples but not most of the examples. However, recall that one of the main objectives of local pattern detection is to find unknown facts in the domain. Our undirected exception rule discovery is considered to be promising since it seeks for various kinds of combinations of a global model and a local pattern. We believe that various combinations of representations for a global model and its local patterns will be proposed in the coming years, which would results in proliferation of local pattern detection research. Our endeavor can be considered as an early attempt in this research stream.

Acknowledgement

This work was partially supported by the grant-in-aid for scientific research on priority area "Active Mining" from the Japanese Ministry of Education, Culture, Sports, Science and Technology.

References

[1] R. Agrawal, H. Mannila, R. Srikant, H. Toivonen, and A. I. Verkamo, "Fast Discovery of Association Rules", *Advances in Knowledge Discovery and Data Mining*, eds. U. M. Fayyad, G. Piatetsky-Shapiro, P. Smyth, and R. Uthurusamy, AAAI/MIT Press, Menlo Park, Calif., 1996, pp. 307–328.

[2] Dagstuhl 2004, Dagstuhl seminar: Detecting Local Patterns, http://www-ai.cs.uni-dortmund.de/DAGSTUHL2004/index.html, 2004 (current November 18, 2004).

[3] J. Dougherty, R. Kohavi, and M. Sahami, "Supervised and Unsupervised Discretization of Continuous Features", *Proc. Twelfth Int'l Conf. Machine Learning (ICML)*, Morgan Kaufmann, San Francisco, 1995, pp. 194–202.

[4] F. Hussain, H. Liu, E. Suzuki, and H. Lu, "Exception Rule Mining with a Relative Interestingness Measure", *Knowledge Discovery and Data Mining, LNAI 1805 (PAKDD)*, Springer, Berlin, 2000, pp. 86–97.

[5] W. Klösgen, "Explora: A Multipattern and Multistrategy Discovery Approach", *Advances in Knowledge Discovery and Data Mining*, eds. U. M. Fayyad *et al.*, AAAI/MIT Press, Menlo Park, Calif., 1996, pp. 249–271.

[6] B. Liu, W. Hsu, L.F. Mun, and H.Y. Lee, "Finding Interesting Patterns Using User Expectations", *IEEE Trans. Knowledge and Data Eng.*, **11**, 1999, pp. 817–832.

[7] B. Liu, W. Hsu, and Y. Ma, "Pruning and Summarizing the Discovered Associations", *Proc. Fifth ACM SIGKDD Int'l Conf. Knowledge Discovery and Data Mining (KDD)*, 1999, pp. 125–134.

[8] B. Padmanabhan and A. Tuzhilin, "A Belief-Driven Method for Discovering Unexpected Patterns", *Proc. Fourth Int'l Conf. Knowledge Discovery and Data Mining (KDD)*, AAAI Press, Menlo Park, Calif., 1998, pp. 94–100.

[9] A. Silberschatz and A. Tuzhilin, "What Makes Patterns Interesting in Knowledge Discovery Systems", *IEEE Trans. Knowledge and Data Eng.*, **8**, 1996, pp. 970–974.

[10] P. Smyth and R. M. Goodman, "An Information Theoretic Approach to Rule Induction from Databases", *IEEE Trans. Knowledge and Data Eng.*, **4**, 1992, pp. 301–316.

[11] E. Suzuki and M. Shimura, "Exceptional Knowledge Discovery in Databases Based on Information Theory", *Proc. Second Int'l Conf. Knowledge Discovery and Data Mining (KDD)*, AAAI Press, Menlo Park, Calif., 1996, pp. 275–278.

[12] E. Suzuki, "Discovering Unexpected Exceptions: A Stochastic Approach", *Proc. Fourth International Workshop on Rough Sets, Fuzzy Sets, and Machine Discovery (RSFD)*, 1996, pp. 225–232.

[13] E. Suzuki, "Autonomous Discovery of Reliable Exception Rules", *Proc. Third Int'l Conf. Knowledge Discovery and Data Mining (KDD)*, AAAI Press, Menlo Park, Calif., 1997, pp. 259–262.

[14] E. Suzuki and Y. Kodratoff, Discovery of Surprising Exception Rules Based on Intensity of Implication, Principles of Data Mining and Knowledge Discovery, LNAI 1510 (PKDD), Springer, 1998, pp. 10–18.

[15] E. Suzuki, "Scheduled Discovery of Exception Rules", *Discovery Science, LNAI 1721 (DS)*, Springer, Berlin, 1999, pp. 184–195.

[16] E. Suzuki and S. Tsumoto, "Evaluating Hypothesis-Driven Exception-Rule Discovery with Medical Data Sets", *Knowledge Discovery and Data Mining, LNAI 1805 (PAKDD)*, Springer, Berlin, 2000, pp. 208–211.

[17] E. Suzuki and S. Tsumoto, Evaluating Hypothesis-driven Exception-rule Discovery with Medical Data Sets, Knowledge Discovery and Data Mining, LNAI 1805 (PAKDD), Springer, 2000, pp. 208–211.

[18] E. Suzuki, "Mining Bacterial Test Data with Scheduled Discovery of Exception Rules", *Proc. Int'l Workshop of KDD Challenge on Real-world Data (KDD Challenge)*, Kyoto, Japan, 2000, pp. 34–40.

[19] E. Suzuki and J.M. Żytkow, Unified Algorithm for Undirected Discovery of Exception Rules, Principles of Data Mining and Knowledge Discovery, LNAI 1910 (PKDD), Springer, 2000, pp. 169–180.

[20] E. Suzuki: "Undirected Discovery of Interesting Exception Rules", International Journal of Pattern Recognition and Artificial Intelligence, Vol. 16, No. 8, 2002, pp. 1065–1086.

[21] E. Suzuki: "Evaluation Scheme for Exception Rule/Group Discovery", Intelligent Technologies for Information Analysis, Springer, Berlin, 2000, pp. 89–108.

[22] E. Suzuki: "Unified Algorithm for Undirected Discovery of Exception Rules", International Journal of Intelligent Systems (accepted for publication).

[23] S. Tsumoto *et al.*, "Comparison of Data Mining Methods using Common Medical Datasets", *ISM Symp.: Data Mining and Knowledge Discovery in Data Science*, 1999, pp. 63–72.

[24] N. Yugami, Y. Ohta, and S. Okamoto, "Fast Discovery of Interesting Rules", *Knowledge Discovery and Data Mining, LNAI 1805 (PAKDD)*, Springer, Berlin, 2000, pp. 17–28.

From Local to Global Analysis of Music Time Series

Claus Weihs and Uwe Ligges

Fachbereich Statistik, Universität Dortmund, D-44221 Dortmund, Germany*

Abstract. Local and more and more global musical structure is analyzed from audio time series by time-series-event analysis with the aim of automatic sheet music production and comparison of singers. Note events are determined and classified based on local spectra, and rules of bar events are identified based on accentuation events related to local energy. In order to compare the performances of different singers global summary measures are defined characterizing the overall performance.

1 Introduction

Music has obviously a global structure. At least classical music is played from well-structured scores. Music has, however, also local structures, the most local structure being a period of time with a certain frequency, the most local structure relevant for scores is a note. Obvious more global structures are measures, indicated by bars, and musical motifs, phrases, etc. Such a hierarchy of more and more global structure might be revealed by means of automatic analysis of music time series. Such analysis is demonstrated by means of transcription of vocal time series into sheet music. With the performance of songs, however, more global structure can be identified. Apart from pitch correctness especially timbre gives a basis for comparison of different singers. For such comparison global characteristics of performances are derived.

The basic data was generated by an experiment where 17 singers, amateurs as well as professionals, all voice types, sung the classical song "Tochter Zion" (G.F. Händel), the piano accompaniment played back via headphones (Weihs et al., 2001). The transcription of these performances to sheet music was carried out by the analysis of the corresponding time series followed by classification using minimal background information about the piece of music and the singer in order to be able to automatically transcribe unknown music as well (Weihs and Ligges, 2003a).

The analysis is embedded in a more general concept of combination of time series and event analysis (cp. Figure 1; Morik, 2000). Events are derived from time series, event rules are derived from events, and time series models might be directly derived from time series or from event rules. The adequacy of this general

---

* The financial support of the Deutsche Forschungsgemeinschaft (SFB 475, "Reduction of complexity in multivariate data structures") is gratefully acknowledged.

K. Morik et al. (Eds.): Local Pattern Detection, LNAI 3539, pp. 217–231, 2005.

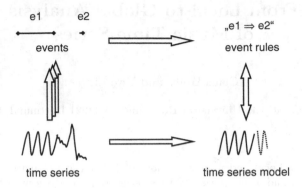

Fig. 1. Time-Series-Event Diagram

scheme for the indicated analysis of music time series is discussed. Steps from local to global analysis of music time series concerning automatic production of sheet music are:

1. Identification of musical local structure by testing against background in local blocks of the time series.
2. Pitch estimation in local blocks.
3. Identification of tone change events on the basis of pitch estimation, after smoothing off vibrato.
4. Classification of notes (**note events**) corresponding to different tones on the basis of constant tempo (**static quantization**).
5. Identification of local tempo, re-classification of notes (**dynamic quantization**).
6. Combination of notes to measures (**bar events**) by identification of meter via identification of high relative energy.
7. **Key identification** by comparing the identified notes with notes expected in keys.
8. Identification of **rhythm** by comparison with rhythm patterns.
9. Combination of notes to **motifs** by identification of repeated similar series of notes.
10. Combination of bars to **phrases** (e.g. 2 bars).
11. ...

For comparison of the performances of different singers characterizations of the performance of the song as a whole are defined, again based on the spectra of local blocks (Weihs and Ligges, 2003b).

Section 2 introduces the data the example analysis is based upon. Section 3 introduces blocking, the basic data preparation for finding local structure in music time series. Section 4 discusses automatic transcription into sheet music, and section 5 global comparison of singers. Section 6 gives a conclusion.

Fig. 2. Sheet Music of "Tochter Zion" (G.F. Händel)

2   Data

The sheet music of "Tochter Zion" (G.F. Händel) can be found in Figure 2. Note the "ABA" structure of the song. This song was sung by 17 singers and recorded in CD-quality (44100 Hz, 16 bit), but down sampled to 11025 Hz before use. Depending on the task, the corresponding time series, so-called wave, was transformed to spectra, e.g. for pitch estimation, and to energy, e.g. for tempo or meter analysis, both locally in blocks of 512 observations (see next section). For global comparisons of performances spectral characterizations of whole performances are derived from local spectra of 2048 observations. Typical waves and corresponding periodograms, here for the syllable "Zi" (c'' with 523.25 Hz), of an amateur and a professional singer look as in Figure 3, and Figure 4.

Energy is generated from the wave observations w_i by means of the formula:

$$\text{energy} = 20 \cdot \log_{10} \sum_{i=1}^{n} |w_i|, \tag{1}$$

with block size $n = 512$. Local energy is analyzed for the accompaniment (see Figure 5).

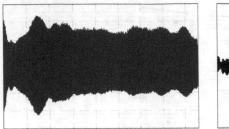

Fig. 3. Waves for syllable "Zi" (c'' with 523.25 Hz), amateur and professional

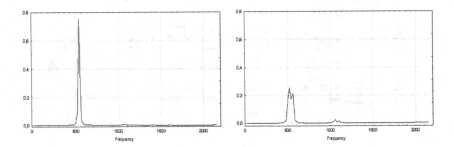

Fig. 4. Periodograms for syllable "Zi" (c'' with 523.25 Hz), amateur and professional

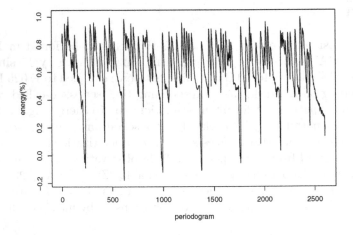

Fig. 5. Local energy of accompaniment

3   Local Analysis and Local Structure

Let us continue with some general arguments about local structure corresponding to local events in time series. Musical local structure can be distinguished from background by testing against zero energy and noise in local blocks of the time series. Thus, in our analysis locality is related to time, and relevance of structure is determined by signal amplitude and its harmonics in contrast to noise. In each local block the signal is tested whether its energy is nonzero and whether its spectrum can be distinguished from spectra typical for noise. Only after musical structure is detected this way, pitch is estimated and more global structure is looked for.

Here is our procedure of testing against background:

1. Test peaks in spectrum against zero amplitude by checking whether relative peak height is high, i.e. bigger than 1% of maximum amplitude, and
2. test for noise by checking whether in the 20% blocks with lowest energy there are blocks with more than 10 high peaks.

Thus low energy blocks with a large number of peaks in the spectrum, indicating noise, are identified as background. Note that harmonic structure can be identified by a small number of peaks in the spectrum which are strongly related to each other in that beneath the fundamental frequency only multiples appear in the spectrum, the so-called overtones. The parameters in our testing procedure, namely the percentages and the number 10 of high peaks were fixed empirically and might be improved by optimization, even individually for a singer.

In order to be able to identify local events, one has to analyze the time series in a granularity that allows for identification of relevant events. Thus, the first task in the analysis is the identification of the size of blocks of observations in the time series to be analyzed. On the one hand, the smaller the blocks, the more exact the time period of an event can be identified. On the other hand, the smaller the number of observations in such a block, the more uncertain is the information on the event in the block. Moreover, it should be clear that event information has to be somewhat redundant to be able to be sufficiently certain about an event. For time series event analysis this can be interpreted as the requirement for 'enough' blocks 'supporting' the event. Obviously, however, more blocks will lead to smaller blocks and to more uncertain information. Overall, block size is a very important topic to be decided upon in the beginning of time-series-event analysis.

In our application, the most basic events are related to notes, which, again, are related to frequencies of signals. Frequencies are best derived from spectral densities. Spectral densities, however, are only observed at Fourier frequencies. The distance of Fourier frequencies is determined by the analyzed block size. E.g., in the case of a sampling rate of 11025 observations per second and a block size of 512 observations, the distance of Fourier frequencies is $11025/512 = 21.5$ Hz. If frequency estimates would be restricted to Fourier frequencies, then this distance would determine the precision of estimates. This would lead to unacceptably large time series blocks. If neighboring Fourier frequencies are too distant especially for identifying low tones, e.g. of a bass singer, it appears to be necessary to estimate the frequency of the realized tone between the observed frequencies. Our idea is to estimate the maximum of a quadratic model fitted to the Fourier frequency with maximum mass and its left and right neighbor. This leads to pitch estimated by interpolated peaking Fourier frequencies:

$$\text{pitch} := h + ((s - h)/2)\sqrt{ds/dh}, \tag{2}$$

where h = peaking Fourier frequency, s = peaking neighbor, dh, and ds are the corresponding density values. The quality of this method is tested with Midi tones corresponding to the tones sung by human voices. This resulted in a very acceptable maximum error lower than 2 Hz. Moreover, we use half overlapping blocks in our analysis. This leads to 12 blocks corresponding to an eighth for our application if constant tempo is assumed. This was assumed to be enough information for note identification for eighths which are the shortest note appearing in the analyzed song. Based on these arguments, a block size of 512 observations is used in the further analysis.

4　Transcription

Transcription of waves to sheet music can be divided into at least 5 steps:

- Separation of a single voice from other sound,
- segmentation of the sound of the selected voice into segments corresponding to notes, silence or noise,
- quantization, i.e. the derivation of relative lengths of notes,
- meter detection in order to separate notes by bars,
- key determination and
- final transcription into sheet music.

In our project, separation was already carried out by recording, i.e. the singing voice and the piano accompaniment were separated to different channels. Hyvärinen et al. (2001) propose *ICA* for polyphonic sound separation. See von Ameln (2001) for a music example. Segmentation was carried out by pitch estimation followed by classification to a corresponding note.

In our project, segmentation into notes is based on the pitch estimation described in the previous section. The segmentation procedure is described in detail in the next subsection. For alternatives in the literature cp., e.g., Cano et al. (1999) describing a Hidden Markov Model, and Dixon (1996) proposing a method using direct pitch estimation.

In quantization the relative lengths of notes (eighth notes, quarter notes, etc.) are derived from estimated absolute lengths. For this, global or local tempo is derived from accompaniment. In our project the sound is separated into eighths first, since we can assume that an eighth is the shortest note. For an alternative see Cemgil et al. (2000).

Meter identification is carried out by comparing the pattern of accentuation, i.e. the peaking energy distances, to standards corresponding to 4/4 and 3/4 (only, at the moment). Compare also Klapuri (2003). Key detection is postponed to the future, and final transcription into music notation is carried out by an

Interface (Preusser et al., 2002) from R (R Development Core Team, 2004) to LilyPond (Nienhuys et al., 2002). This final step also comprises the combination of eighths of equal pitch to longer notes.

4.1 Segmentation Procedure

The task of segmentation is to identify so-called note events from the music time series corresponding to one voice. The procedure can be divided into the following steps:

- Passing through the vocal time series by sections of given size n ($n = 512$ appeared to be appropriate for a wave file sampled with 11kHz).
- Pitch estimation for each section by estimation of the spectral density by means of a periodogram, and the interpolation described above of frequencies of highest periodogram peaks.
- Note classification using estimated fundamental frequencies and the corresponding overtones, given the frequency of diapason a', which might have been estimated.
- Smoothing of classified notes because of vibrato. In our project a doubled *running median* with window width 9 is used. For an alternative vibrato analysis see, e.g., Rossignol et al. (1999).
- Segmentation, iff a change in the smoothed list of notes occurs.

4.2 Transcription by Example

Transcription is now demonstrated by means of an example: the last part A of the ABA scheme of "Tochter Zion" sung by a professional soprano singer. The pitch of each 512-section of the vocal time series is estimated on the basis of the estimated frequency of diapason a' of accompaniment equal to 443.5 Hz, and each corresponding note is classified.

In Table 1 the raw classified sections of the first measure are given, where 0 corresponds to diapason a', other integers represent the distance of halftones from a', and silence and quiet noise is represented by NA. The singer has an intensive vibrato: classification switches rapidly between 2 (b'), 3 (c''), and 4 ($c\#''$) in the first 2 rows (changes marked by *). Smoothing does not smooth off the intensive vibrato completely (see Table 2), the second half of the note is classified one halftone flat. And moreover, the first sections are classified as c''' instead of c'' since only the first overtone is appearing in the spectrum (see Figure 6).

In Figure 7 a first impression of the sheet music is given. The line indicates the classified note events after smoothing without quantization. The indicated bars are just showed for orientation. The progression of corresponding energy is shown as well, low energy reflecting breathing, silence, and strong consonants. Such parts will not be counted as errors in the following.

For meter assessment accentuation events are derived from accompaniment. By smoothing constant energy in quarters is produced (cp. Figure 8), assuming

Table 1. Raw classified sections of the first bar

```
          *     *           * *         *           *
NA NA -12 NA   2  2 15 15 15 15 15 15  2 3 3 3 3 3 2 2   2  2  4  4
 3  2   2  2   2  4  4  3  2  2  2  2  3 3 2 2 2 2 2 2 -30 NA NA NA
NA NA  NA -27 -14  0  0 -1  0  0  0  0  0 0 0 0 0 0 0 0   0 -1 -1 -1
 0  0   0  0  -1 -1 -1  0  0  0  0  0 -1 0 1 1 1 1 1 1   1  0 NA NA
```

Table 2. Smoothed classified sections of the first bar

```
15 15 15 15 15 15 15 15 15 15 15 15  3  3  3  3  3  3  3  3  3  2  2  2
 2  2  2  2  2  2  2  2  2  2  2  2  2  2  2  2  2  2  2  2  2  2 NA NA NA
NA NA NA  0  0  0  0  0  0  0  0  0  0  0  0  0  0  0  0  0  0  0  0  0
 0  0  0  0  0  0  0  0  0  0  0  0  0  0  0  1  1  1  1  1  1  1  1 NA NA
```

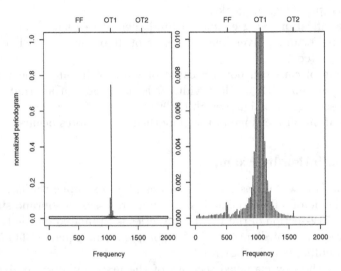

Fig. 6. c'' estimation one octave to high: standard periodogram, and zoomed

global tempo (see below). Then, an *accentuation event* A_i is defined as
a turning point quarter with energy(quarter) > 0.75,
i.e. such a quarter is a turning point in that the energy of neighboring quarters is
lower, and its energy is high peaking. Then $D_{i-1} := \#(A_i) - \#(A_{i-1})$ indicates
the *number of quarters between accentuation events*, $\#(A_i)$ being the running
quarter number of accentuation event A_i. This can be used to establish rules for
the different meters, e.g.

$$\text{no.}(D_{i-1} = 4) > \text{no.}(D_{i-1} = 3) \quad \Rightarrow \quad 4/4 \text{ meter},\tag{3}$$

meaning that differences of 4 are appearing more often than differences of 3.
    In the next step to sheet music, static quantization is carried out, assuming
unit = eighth, and no.(eighths) are known. Global tempo is then characterized

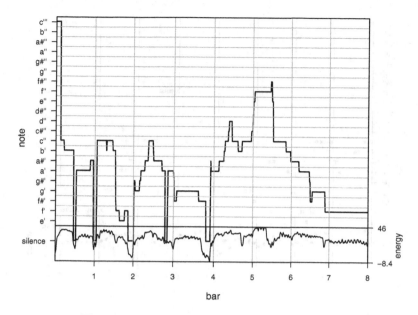

Fig. 7. Progression of note events and energy

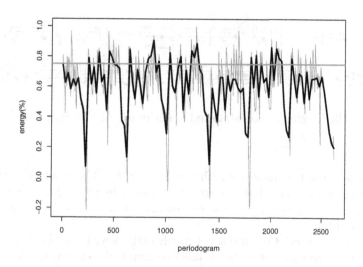

Fig. 8. Smoothed energy indicating meter

by length(eighth) = length(series)/no.(eighths). Note events are now related to eighths, and re-classified as the statistical mode of the 12 classified sections of each eighth note in Table 2. Note that each row of Table 2 corresponds to the 24 blocks of one quarter, i.e. two succeeding eighths. Since 4/4 meter was identified, bar events are placed after each 8 eighths, assuming to be known that

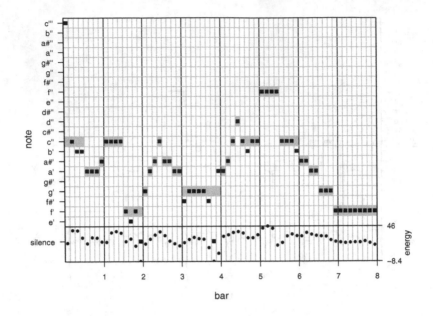

Fig. 9. Progression of notes after quantization

Fig. 10. Original, and estimated sheet music of "Tochter Zion"

singing starts in the first measure with the first eighth. The result of this static quantization together with bar placement is shown in the Figure 9 which can "directly" be transcribed into notes. For comparison, true notes are shown as grey horizontal bars.

For the final transcription into sheet music eighths with equal pitch are combined and music symbols for *rests* are used to transcribe silence and low energy noise. The result (cp. Figure 10) is judged by error rates calculated as follows:

$$\text{error rate} := \frac{\text{no.(erroneously classified eighth notes, without counting rests)}}{\text{no.(all eighth notes)} - \text{no.(eighth rests)}}$$

Note that in our example there are 64 eighth notes in 8 bars to be classified. Obviously, there are 9 erroneously classified eighth notes, and 2 eighth rests, thus the error rate is $9/62 = 15\%$. The transcriptions of the other singers' performances gave error rates from 4% to 26%. Table 3 shows a comparison for

Table 3. Comparing error rates with and without counting rests

singer	overall error (counting rests as errors)	error (cleared for rests)
1	0.11	0.07
2	0.19	0.04
3	0.25	0.21
4	0.23	0.20
5	0.11	0.07
6	0.16	0.10
7	0.30	0.26

4 soprano and 3 tenor singers of overall error rates (rests have been counted as errors, even if the singer was not singing) and the error rates cleared for rests as described above.

To assess our outcomes, one should, on the one hand, recognize that we used some a-priori information like knowledge about the shortest note length (eighth note, important for smoothers), the overall number of eighths (for global tempo), and the begin of first sung eighth in the first bar (for bar setting). On the other hand, one should note that the error rate is the sum of various kinds of errors:

- errors of the transcription algorithm, but also
- errors of the singer's performances,
- esp. errors from inaccurate timings of singer,
- errors from static quantization: local tempo of accompaniment was ignored.

This might lead to an analysis of local tempo of the singer or of the accompaniment. At least in our experiment, however, local tempo of singers was very irregular, and local tempo of accompaniment was only followed very roughly by the singer so that utilizing of local tempo did not improve quantization.

In the future we will try to overcome the need of a-priori information, and moreover, we will try to improve our transcription by modelling of vibrato aiming at an improvement of the note classification.

5 Global Analysis: Comparison of Singers

Up to now we have analyzed pitch related information in the audio time series locally and more and more globally. This way, we were aiming at automatic transcription into scores. What we have nearly totally ignored until now is the fact that different singers produce different performances of the song. In order to compare such performances it is necessary to define global summary measures characterizing the overall performance. One possible such measure is the size of pitch errors compared to the given score to be reproduced. A possible ratio scale is 'parts of half tone' pht. After assessment of pitch correctness by pht we concentrate on the information in the spectrum residual to pitch information. This

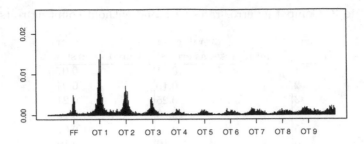

Fig. 11. Pitch independent periodogram

is realized by elimination of pitch from the spectrum. The Fourier Frequencies are linearly rescaled, so that the frequency corresponding to the fundamental is mapped to 1, the frequencies corresponding to the first overtones are mapped to 2, 3, etc. Overlaying and averaging the spectra of different blocks lead to, what we call, the pitch independent spectrum (cp. Figure 11). This time we used half overlaying blocks of $n = 2048$ observations as a basis for periodograms. In pitch independent spectra the size and the shape of the first 13 partials, i.e. the fundamental frequency (FF) and the first 12 overtones (OT1, ..., OT12), are used as characterizations of the residual information in the spectrum after pitch elimination, which is said to be related to the individual timbre of the performance. In order to measure the size of the peaks in the spectrum, the mass (weight) of the peaks of the partials are determined as the sum of the percentage shares of those parts of the corresponding peak in the spectrum which are higher than a pre-specified threshold. The shape of a peak cannot easily be described. Therefore, we only use one simple characteristic of the shape, namely the width of the peak of the partials. The width of a peak is measured by the half tone distance between the smallest and the biggest frequency of the peak with a spectral height above a pre-specified threshold. Mass is measured as a percentage (%), whereas width is measured in parts of halftones (pht). For details on the computation of the measures see Güttner (2001). Based on music theory as a last voice characteristic formant intensity is chosen. This gives the part of mass lying in what one calls the singer's formant lying between 2000 and 3500 Hz individual for the voice types (Soprano, Alto, Tenor, Bass). A large singer's formant characterizes the ability to dominate an orchestra. Overall, every singer is characterized by the above 28 characteristics as a basis for comparison. Figure 12 illustrates the voice print corresponding to the whole song "Tochter Zion" for a particular singer. For masses and widths boxplots are indicating variation over the involved tones.

As an example for comparison let us consider the masses of professional and amateur bass and soprano singers. Figure 13 illustrates that professional singers (Becker and Hasse) have less mass at the fundamental frequency and more mass on higher overtones. The latter is especially true the professional bass singer who has particularly large mass at the singer's formant (cp. Figure 14). For sopranos the singer's formant does not appear to be that pronounced in general.

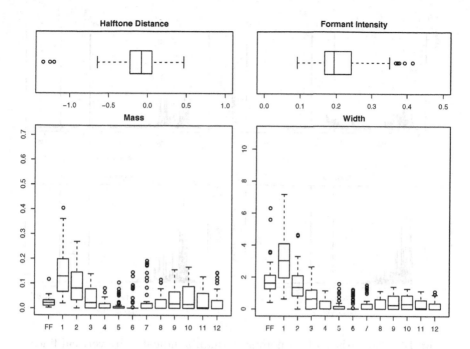

Fig. 12. Voice print: Professional Bass Singer

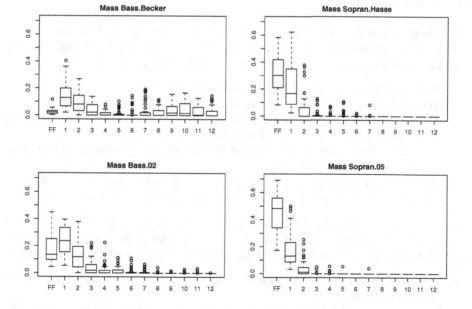

Fig. 13. Voice prints: Comparison of Masses

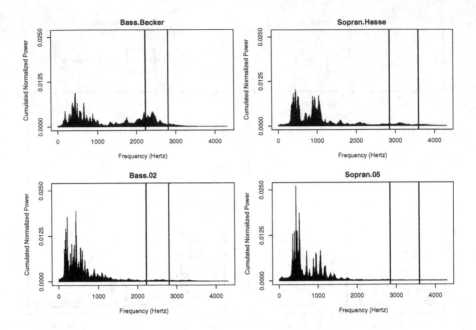

Fig. 14. Comparison of Formants: Formants indicated by vertical lines

6   Conclusion

Our analysis was embedded in a more general concept of combination of time series and event analysis (cp. Morik, 2000). We derived three different kinds of events, for other events event rules have to be derived in the future:

- **Note events** are derived by pitch analysis of time series.
- **Accentuation events** are derived by energy analysis of time series.
- For **bar events** an **event rule** was derived by comparison of accentuation events with meter related accentuation patterns.
- **Rhythm rules** have to be derived in the future by comparison of note lengths with rhythm patterns.
- **Higher structuring rules** for identification of motifs, and phrases of music pieces have to be derived in the future from prescribed event rule types.

Note that we did not derive time series models corresponding to event rules to complete the time-series-event analysis. Concerning local and global analysis we investigated the local and somewhat more global structure of a piece of music:

- Very local pitch and energy estimation was the basis for identification of more global note events.
- Note events were basic for transcription: Identification was based on smoothing of local preliminary notes.
- Local energy estimation of quarters was the basis for accentuation events and meter identification.

– Bar events structure note events: Identification was based on accentuation events.

More global analyses would be key analysis, rhythm analysis, and motif and phrase analysis of music pieces. This was postponed to future research.

For the global comparison of different singers voice prints were developed. In particular, this lead to the identification of pitch independent spectral differences of basses and sopranos and of professionals and amateurs.

References

[1] von Ameln, F.: Blind source separation in der Praxis. Diploma Thesis, Fachbereich Statistik, Universität Dortmund, Germany (2001)

[2] Cano, P., Loscos, A., Bonada, J.: Score-Performance Matching using HMMs. In: Proceedings of the International Computer Music Conference, Beijing, China (1999)

[3] Cemgil, T., Desain, P., Kappen, B.: Rhythm Quantization for Transcription. Computer Music Journal 24 (2000) 60–76

[4] Dixon, S.: Multiphonic Note Identification. Australian Computer Science Communications 17 (1996) 318–323

[5] Güttner, J. : Klassifikation von Gesangsdarbietungen. Diploma Thesis, Fachbereich Statistik, Universität Dortmund, Germany (2001)

[6] Hyvärinen, A., Karhunen, J., Oja, E.: Independent Component Analysis. John Wiley and Sons, New York (2001)

[7] Klapuri, A.: Automatic Transcription of Music. In: Proceedings of the Stockholm Music Acoustics Conference, SMAC03 (2003)

[8] Morik, K.: The Representation Race – Preprocessing for Handling Time Phenomena. In López de Mántaras, R., Plaza, E., eds.: Proceedings of the European Conference on Machine Learning 2000 (ECML 2000), Lecture Notes in Artificial Intelligence 1810, Berlin, Springer (2000)

[9] Nienhuys, H.W., Nieuwenhuizen, J., et al.: GNU LilyPond – The Music Typesetter. Free Software Foundation. (2002) version 1.6.5

[10] Preusser, A., Ligges, U., Weihs, C.: Ein R Exportfilter für das Notations- und Midi-Programm LilyPond. Arbeitsbericht 35, Fachbereich Statistik, Universität Dortmund, Germany (2002)

[11] R Development Core Team: R: A Language and Environment for Statistical Computing. R Foundation for Statistical Computing, Vienna, Austria (2004)

[12] Rossignol, S., Depalle, P., Soumagne, J., Rodet, X., Collette, J.L.: Vibrato: Detection, Estimation, Extractiom, Modification. In: Proceedings 99 Digital Audio Effects Workshop (1999)

[13] Weihs, C., Berghoff, S., Hasse-Becker, P., Ligges, U.: Assessment of Purity of Intonation in Singing Presentations by Discriminant Analysis. In Kunert, J., Trenkler, G., eds.: Mathematical Statistics and Biometrical Applications, Lohmar, Josef Eul Verlag (2001) 395–410

[14] Weihs, C., Ligges, U.: Automatic transcription of singing performances. In: Bulletin of the International Statistical institute, 54th Session, Proceedings. Volume LX. (2003a) 507–510

[15] Weihs, C., Ligges, U.: Voice Prints as a Tool for Automatic Classification of Vocal Performance. In: Kopiez, R., Lehmann, A.C., Wolther, I., Wolf, C., eds.: Proceedings of the 5th Triennial ESCOM Conference, Hanover University of Music and Drama, Germany, 8-13 September 2003 (2003b) 332–335

Author Index

Adams, Niall M., 39

Baron, Steffan, 190
Bonchi, Francesco, 1
Boulicaut, Jean-François, 115

Džeroski, Sašo, 71

Fürnkranz, Johannes, 20

Giannotti, Fosca, 1
Grobelnik, Marko, 89

Hand, David J., 39
Heard, Nick A., 39
Höppner, Frank, 53

Köpcke, Hanna, 98

Lavrač, Nada, 71
Ligges, Uwe, 217

Mladenic, Dunja, 89
Morik, Katharina, 98

Pensa, Ruggero G., 115

Radvanyi, Francois, 135
Rouveirol, Céline, 135
Rüping, Stefan, 153

Scholz, Martin, 171
Spiliopoulou, Myra, 190
Suzuki, Einoshin, 207

Weihs, Claus, 217

Železný, Filip, 71

Lecture Notes in Artificial Intelligence (LNAI)

Vol. 3626: B. Ganter, G. Stumme, R. Wille (Eds.), Formal Concept Analysis. X, 349 pages. 2005.

Vol. 3596: F. Dau, M.-L. Mugnier, G. Stumme (Eds.), Conceptual Structures: Common Semantics for Sharing Knowledge. XI, 467 pages. 2005.

Vol. 3587: P. Perner, A. Imiya (Eds.), Machine Learning and Data Mining in Pattern Recognition. XVII, 695 pages. 2005.

Vol. 3584: X. Li, S. Wang, Z.Y. Dong (Eds.), Advanced Data Mining and Applications. XIX, 835 pages. 2005.

Vol. 3575: S. Wermter, G. Palm, M. Elshaw (Eds.), Biomimetic Neural Learning for Intelligent Robots. IX, 383 pages. 2005.

Vol. 3571: L. Godo (Ed.), Symbolic and Quantitative Approaches to Reasoning with Uncertainty. XVI, 1028 pages. 2005.

Vol. 3559: P. Auer, R. Meir (Eds.), Learning Theory. XI, 692 pages. 2005.

Vol. 3558: V. Torra, Y. Narukawa, S. Miyamoto (Eds.), Modeling Decisions for Artificial Intelligence. XII, 470 pages. 2005.

Vol. 3554: A. Dey, B. Kokinov, D. Leake, R. Turner (Eds.), Modeling and Using Context. XIV, 572 pages. 2005.

Vol. 3539: K. Morik, J.-F. Boulicaut, A. Siebes (Eds.), Local Pattern Detection. XI, 233 pages. 2005.

Vol. 3538: L. Ardissono, P. Brna, A. Mitrovic (Eds.), User Modeling 2005. XVI, 533 pages. 2005.

Vol. 3533: M. Ali, F. Esposito (Eds.), Innovations in Applied Artificial Intelligence. XX, 858 pages. 2005.

Vol. 3528: P.S. Szczepaniak, J. Kacprzyk, A. Niewiadomski (Eds.), Advances in Web Intelligence. XVII, 513 pages. 2005.

Vol. 3518: T.B. Ho, D. Cheung, H. Liu (Eds.), Advances in Knowledge Discovery and Data Mining. XXI, 864 pages. 2005.

Vol. 3508: P. Bresciani, P. Giorgini, B. Henderson-Sellers, G. Low, M. Winikoff (Eds.), Agent-Oriented Information Systems II. X, 227 pages. 2005.

Vol. 3505: V. Gorodetsky, J. Liu, V. A. Skormin (Eds.), Autonomous Intelligent Systems: Agents and Data Mining. XIII, 303 pages. 2005.

Vol. 3501: B. Kégl, G. Lapalme (Eds.), Advances in Artificial Intelligence. XV, 458 pages. 2005.

Vol. 3492: P. Blache, E. Stabler, J. Busquets, R. Moot (Eds.), Logical Aspects of Computational Linguistics. X, 363 pages. 2005.

Vol. 3488: M.-S. Hacid, N.V. Murray, Z.W. Raś, S. Tsumoto (Eds.), Foundations of Intelligent Systems. XIII, 700 pages. 2005.

Vol. 3476: J. Leite, A. Omicini, P. Torroni, P. Yolum (Eds.), Declarative Agent Languages and Technologies II. XII, 289 pages. 2005.

Vol. 3464: S.A. Brueckner, G.D.M. Serugendo, A. Karageorgos, R. Nagpal (Eds.), Engineering Self-Organising Systems. XIII, 299 pages. 2005.

Vol. 3452: F. Baader, A. Voronkov (Eds.), Logic for Programming, Artificial Intelligence, and Reasoning. XI, 562 pages. 2005.

Vol. 3451: M.-P. Gleizes, A. Omicini, F. Zambonelli (Eds.), Engineering Societies in the Agents World. XIII, 349 pages. 2005.

Vol. 3446: T. Ishida, L. Gasser, H. Nakashima (Eds.), Massively Multi-Agent Systems I. XI, 349 pages. 2005.

Vol. 3445: G. Chollet, A. Esposito, M. Faundez-Zanuy, M. Marinaro (Eds.), Nonlinear Speech Modeling and Applications. XIII, 433 pages. 2005.

Vol. 3438: H. Christiansen, P.R. Skadhauge, J. Villadsen (Eds.), Constraint Solving and Language Processing. VIII, 205 pages. 2005.

Vol. 3430: S. Tsumoto, T. Yamaguchi, M. Numao, H. Motoda (Eds.), Active Mining. XII, 349 pages. 2005.

Vol. 3419: B. Faltings, A. Petcu, F. Fages, F. Rossi (Eds.), Constraint Satisfaction and Constraint Logic Programming. X, 217 pages. 2005.

Vol. 3416: M. Böhlen, J. Gamper, W. Polasek, M.A. Wimmer (Eds.), E-Government: Towards Electronic Democracy. XIII, 311 pages. 2005.

Vol. 3415: P. Davidsson, B. Logan, K. Takadama (Eds.), Multi-Agent and Multi-Agent-Based Simulation. X, 265 pages. 2005.

Vol. 3403: B. Ganter, R. Godin (Eds.), Formal Concept Analysis. XI, 419 pages. 2005.

Vol. 3398: D.-K. Baik (Ed.), Systems Modeling and Simulation: Theory and Applications. XIV, 733 pages. 2005.

Vol. 3397: T.G. Kim (Ed.), Artificial Intelligence and Simulation. XV, 711 pages. 2005.

Vol. 3396: R.M. van Eijk, M.-P. Huget, F. Dignum (Eds.), Agent Communication. X, 261 pages. 2005.

Vol. 3394: D. Kudenko, D. Kazakov, E. Alonso (Eds.), Adaptive Agents and Multi-Agent Systems II. VIII, 313 pages. 2005.

Vol. 3392: D. Seipel, M. Hanus, U. Geske, O. Bartenstein (Eds.), Applications of Declarative Programming and Knowledge Management. X, 309 pages. 2005.

Vol. 3374: D. Weyns, H.V.D. Parunak, F. Michel (Eds.), Environments for Multi-Agent Systems. X, 279 pages. 2005.

Vol. 3371: M.W. Barley, N. Kasabov (Eds.), Intelligent Agents and Multi-Agent Systems. X, 329 pages. 2005.

Vol. 3369: V.R. Benjamins, P. Casanovas, J. Breuker, A. Gangemi (Eds.), Law and the Semantic Web. XII, 249 pages. 2005.

Vol. 3366: I. Rahwan, P. Moraitis, C. Reed (Eds.), Argumentation in Multi-Agent Systems. XII, 263 pages. 2005.

Vol. 3359: G. Grieser, Y. Tanaka (Eds.), Intuitive Human Interfaces for Organizing and Accessing Intellectual Assets. XIV, 257 pages. 2005.

Vol. 3346: R.H. Bordini, M. Dastani, J. Dix, A.E.F. Seghrouchni (Eds.), Programming Multi-Agent Systems. XIV, 249 pages. 2005.

Vol. 3345: Y. Cai (Ed.), Ambient Intelligence for Scientific Discovery. XII, 311 pages. 2005.

Vol. 3343: C. Freksa, M. Knauff, B. Krieg-Brückner, B. Nebel, T. Barkowsky (Eds.), Spatial Cognition IV. XIII, 519 pages. 2005.

Vol. 3339: G.I. Webb, X. Yu (Eds.), AI 2004: Advances in Artificial Intelligence. XXII, 1272 pages. 2004.

Vol. 3336: D. Karagiannis, U. Reimer (Eds.), Practical Aspects of Knowledge Management. X, 523 pages. 2004.

Vol. 3327: Y. Shi, W. Xu, Z. Chen (Eds.), Data Mining and Knowledge Management. XIII, 263 pages. 2005.

Vol. 3315: C. Lemaître, C.A. Reyes, J.A. González (Eds.), Advances in Artificial Intelligence – IBERAMIA 2004. XX, 987 pages. 2004.

Vol. 3303: J.A. López, E. Benfenati, W. Dubitzky (Eds.), Knowledge Exploration in Life Science Informatics. X, 249 pages. 2004.

Vol. 3301: G. Kern-Isberner, W. Rödder, F. Kulmann (Eds.), Conditionals, Information, and Inference. XII, 219 pages. 2005.

Vol. 3276: D. Nardi, M. Riedmiller, C. Sammut, J. Santos-Victor (Eds.), RoboCup 2004: Robot Soccer World Cup VIII. XVIII, 678 pages. 2005.

Vol. 3275: P. Perner (Ed.), Advances in Data Mining. VIII, 173 pages. 2004.

Vol. 3265: R.E. Frederking, K.B. Taylor (Eds.), Machine Translation: From Real Users to Research. XI, 392 pages. 2004.

Vol. 3264: G. Paliouras, Y. Sakakibara (Eds.), Grammatical Inference: Algorithms and Applications. XI, 291 pages. 2004.

Vol. 3259: J. Dix, J. Leite (Eds.), Computational Logic in Multi-Agent Systems. XII, 251 pages. 2004.

Vol. 3257: E. Motta, N.R. Shadbolt, A. Stutt, N. Gibbins (Eds.), Engineering Knowledge in the Age of the Semantic Web. XVII, 517 pages. 2004.

Vol. 3249: B. Buchberger, J.A. Campbell (Eds.), Artificial Intelligence and Symbolic Computation. X, 285 pages. 2004.

Vol. 3248: K.-Y. Su, J. Tsujii, J.-H. Lee, O.Y. Kwong (Eds.), Natural Language Processing – IJCNLP 2004. XVIII, 817 pages. 2005.

Vol. 3245: E. Suzuki, S. Arikawa (Eds.), Discovery Science. XIV, 430 pages. 2004.

Vol. 3244: S. Ben-David, J. Case, A. Maruoka (Eds.), Algorithmic Learning Theory. XIV, 505 pages. 2004.

Vol. 3238: S. Biundo, T. Frühwirth, G. Palm (Eds.), KI 2004: Advances in Artificial Intelligence. XI, 467 pages. 2004.

Vol. 3230: J.L. Vicedo, P. Martínez-Barco, R. Muñoz, M. Saiz Noeda (Eds.), Advances in Natural Language Processing. XII, 488 pages. 2004.

Vol. 3229: J.J. Alferes, J. Leite (Eds.), Logics in Artificial Intelligence. XIV, 744 pages. 2004.

Vol. 3228: M.G. Hinchey, J.L. Rash, W.F. Truszkowski, C.A. Rouff (Eds.), Formal Approaches to Agent-Based Systems. VIII, 290 pages. 2004.

Vol. 3215: M.G.. Negoita, R.J. Howlett, L.C. Jain (Eds.), Knowledge-Based Intelligent Information and Engineering Systems, Part III. LVII, 906 pages. 2004.

Vol. 3214: M.G.. Negoita, R.J. Howlett, L.C. Jain (Eds.), Knowledge-Based Intelligent Information and Engineering Systems, Part II. LVIII, 1302 pages. 2004.

Vol. 3213: M.G.. Negoita, R.J. Howlett, L.C. Jain (Eds.), Knowledge-Based Intelligent Information and Engineering Systems, Part I. LVIII, 1280 pages. 2004.

Vol. 3209: B. Berendt, A. Hotho, D. Mladenic, M. van Someren, M. Spiliopoulou, G. Stumme (Eds.), Web Mining: From Web to Semantic Web. IX, 201 pages. 2004.

Vol. 3206: P. Sojka, I. Kopecek, K. Pala (Eds.), Text, Speech and Dialogue. XIII, 667 pages. 2004.

Vol. 3202: J.-F. Boulicaut, F. Esposito, F. Giannotti, D. Pedreschi (Eds.), Knowledge Discovery in Databases: PKDD 2004. XIX, 560 pages. 2004.

Vol. 3201: J.-F. Boulicaut, F. Esposito, F. Giannotti, D. Pedreschi (Eds.), Machine Learning: ECML 2004. XVIII, 580 pages. 2004.

Vol. 3194: R. Camacho, R. King, A. Srinivasan (Eds.), Inductive Logic Programming. XI, 361 pages. 2004.

Vol. 3192: C. Bussler, D. Fensel (Eds.), Artificial Intelligence: Methodology, Systems, and Applications. XIII, 522 pages. 2004.

Vol. 3191: M. Klusch, S. Ossowski, V. Kashyap, R. Unland (Eds.), Cooperative Information Agents VIII. XI, 303 pages. 2004.

Vol. 3187: G. Lindemann, J. Denzinger, I.J. Timm, R. Unland (Eds.), Multiagent System Technologies. XIII, 341 pages. 2004.

Vol. 3176: O. Bousquet, U. von Luxburg, G. Rätsch (Eds.), Advanced Lectures on Machine Learning. IX, 241 pages. 2004.

Vol. 3171: A.L.C. Bazzan, S. Labidi (Eds.), Advances in Artificial Intelligence – SBIA 2004. XVII, 548 pages. 2004.

Vol. 3159: U. Visser, Intelligent Information Integration for the Semantic Web. XIV, 150 pages. 2004.

Vol. 3157: C. Zhang, H. W. Guesgen, W.K. Yeap (Eds.), PRICAI 2004: Trends in Artificial Intelligence. XX, 1023 pages. 2004.

Vol. 3155: P. Funk, P.A. González Calero (Eds.), Advances in Case-Based Reasoning. XIII, 822 pages. 2004.

Vol. 3139: F. Iida, R. Pfeifer, L. Steels, Y. Kuniyoshi (Eds.), Embodied Artificial Intelligence. IX, 331 pages. 2004.

Vol. 3131: V. Torra, Y. Narukawa (Eds.), Modeling Decisions for Artificial Intelligence. XI, 327 pages. 2004.